T0138081

Lecture Notes in Computer Science　14192

Founding Editors

Gerhard Goos
Juris Hartmanis

The series Lecture Notes in Computer Science (LNCS), including its subseries Lecture Notes in Artificial Intelligence (LNAI) and Lecture Notes in Bioinformatics (LNBI), has established itself as a medium for the publication of new developments in computer science and information technology research, teaching, and education.

LNCS enjoys close cooperation with the computer science R & D community, the series counts many renowned academics among its volume editors and paper authors, and collaborates with prestigious societies. Its mission is to serve this international community by providing an invaluable service, mainly focused on the publication of conference and workshop proceedings and postproceedings. LNCS commenced publication in 1973.

Gernot A. Fink · Rajiv Jain · Koichi Kise ·
Richard Zanibbi
Editors

Document Analysis
and Recognition –
ICDAR 2023

17th International Conference
San José, CA, USA, August 21–26, 2023
Proceedings, Part VI

 Springer

Editors
Gernot A. Fink
TU Dortmund University
Dortmund, Germany

Rajiv Jain
Adobe
College Park, MN, USA

Koichi Kise
Osaka Metropolitan University
Osaka, Japan

Richard Zanibbi
Rochester Institute of Technology
Rochester, NY, USA

ISSN 0302-9743   ISSN 1611-3349 (electronic)
Lecture Notes in Computer Science
ISBN 978-3-031-41730-6   ISBN 978-3-031-41731-3 (eBook)
https://doi.org/10.1007/978-3-031-41731-3

This Springer imprint is published by the registered company Springer Nature Switzerland AG
The registered company address is: Gewerbestrasse 11, 6330 Cham, Switzerland

Foreword

We are delighted to welcome you to the proceedings of ICDAR 2023, the 17th IAPR International Conference on Document Analysis and Recognition, which was held in San Jose, in the heart of Silicon Valley in the United States. With the worst of the pandemic behind us, we hoped that ICDAR 2023 would be a fully in-person event. However, challenges such as difficulties in obtaining visas also necessitated the partial use of hybrid technologies for ICDAR 2023. The oral papers being presented remotely were synchronous to ensure that conference attendees interacted live with the presenters and the limited hybridization still resulted in an enjoyable conference with fruitful interactions.

ICDAR 2023 was the 17th edition of a longstanding conference series sponsored by the International Association of Pattern Recognition (IAPR). It is the premier international event for scientists and practitioners in document analysis and recognition. This field continues to play an important role in transitioning to digital documents. The IAPR-TC 10/11 technical committees endorse the conference. The very first ICDAR was held in St Malo, France in 1991, followed by Tsukuba, Japan (1993), Montreal, Canada (1995), Ulm, Germany (1997), Bangalore, India (1999), Seattle, USA (2001), Edinburgh, UK (2003), Seoul, South Korea (2005), Curitiba, Brazil (2007), Barcelona, Spain (2009), Beijing, China (2011), Washington, DC, USA (2013), Nancy, France (2015), Kyoto, Japan (2017), Sydney, Australia (2019) and Lausanne, Switzerland (2021).

Keeping with its tradition from past years, ICDAR 2023 featured a three-day main conference, including several competitions to challenge the field and a post-conference slate of workshops, tutorials, and a doctoral consortium. The conference was held at the San Jose Marriott on August 21–23, 2023, and the post-conference tracks at the Adobe World Headquarters in San Jose on August 24–26, 2023.

We thank our executive co-chairs, Venu Govindaraju and Tong Sun, for their support and valuable advice in organizing the conference. We are particularly grateful to Tong for her efforts in facilitating the organization of the post-conference in Adobe Headquarters and for Adobe's generous sponsorship.

The highlights of the conference include keynote talks by the recipient of the IAPR/ICDAR Outstanding Achievements Award, and distinguished speakers Marti Hearst, UC Berkeley School of Information; Vlad Morariu, Adobe Research; and Seiichi Uchida, Kyushu University, Japan.

A total of 316 papers were submitted to the main conference (plus 33 papers to the ICDAR-IJDAR journal track), with 53 papers accepted for oral presentation (plus 13 IJDAR track papers) and 101 for poster presentation. We would like to express our deepest gratitude to our Program Committee Chairs, featuring three distinguished researchers from academia, Gernot A. Fink, Koichi Kise, and Richard Zanibbi, and one from industry, Rajiv Jain, who did a phenomenal job in overseeing a comprehensive reviewing process and who worked tirelessly to put together a very thoughtful and interesting technical program for the main conference. We are also very grateful to the members of the Program Committee for their high-quality peer reviews. Thank you to our competition chairs, Kenny Davila, Chris Tensmeyer, and Dimosthenis Karatzas, for overseeing the competitions.

The post-conference featured 8 excellent workshops, four value-filled tutorials, and the doctoral consortium. We would like to thank Mickael Coustaty and Alicia Fornes, the workshop chairs, Elisa Barney-Smith and Laurence Likforman-Sulem, the tutorial chairs, and Jean-Christophe Burie and Andreas Fischer, the doctoral consortium chairs, for their efforts in putting together a wonderful post-conference program.

We would like to thank and acknowledge the hard work put in by our Publication Chairs, Anurag Bhardwaj and Utkarsh Porwal, who worked diligently to compile the camera-ready versions of all the papers and organize the conference proceedings with Springer. Many thanks are also due to our sponsorship, awards, industry, and publicity chairs for their support of the conference.

The organization of this conference was only possible with the tireless behind-the-scenes contributions of our webmaster and tech wizard, Edward Sobczak, and our secretariat, ably managed by Carol Doermann. We convey our heartfelt appreciation for their efforts.

Finally, we would like to thank for their support our many financial sponsors and the conference attendees and authors, for helping make this conference a success. We sincerely hope those who attended had an enjoyable conference, a wonderful stay in San Jose, and fruitful academic exchanges with colleagues.

August 2023   David Doermann
                                                                        Srirangaraj (Ranga) Setlur

Preface

Welcome to the proceedings of the 17th International Conference on Document Analysis and Recognition (ICDAR) 2023. ICDAR is the premier international event for scientists and practitioners involved in document analysis and recognition.

This year, we received 316 conference paper submissions with authors from 42 different countries. In order to create a high-quality scientific program for the conference, we recruited 211 regular and 38 senior program committee (PC) members. Regular PC members provided a total of 913 reviews for the submitted papers (an average of 2.89 per paper). Senior PC members who oversaw the review phase for typically 8 submissions took care of consolidating reviews and suggested paper decisions in their meta-reviews. Based on the information provided in both the reviews and the prepared meta-reviews we PC Chairs then selected 154 submissions (48.7%) for inclusion into the scientific program of ICDAR 2023. From the accepted papers, 53 were selected for oral presentation, and 101 for poster presentation.

In addition to the papers submitted directly to ICDAR 2023, we continued the tradition of teaming up with the International Journal of Document Analysis and Recognition (IJDAR) and organized a special journal track. The journal track submissions underwent the same rigorous review process as regular IJDAR submissions. The ICDAR PC Chairs served as Guest Editors and oversaw the review process. From the 33 manuscripts submitted to the journal track, 13 were accepted and were published in a Special Issue of IJDAR entitled "Advanced Topics of Document Analysis and Recognition." In addition, all papers accepted in the journal track were included as oral presentations in the conference program.

A very prominent topic represented in both the submissions from the journal track as well as in the direct submissions to ICDAR 2023 was handwriting recognition. Therefore, we organized a Special Track on Frontiers in Handwriting Recognition. This also served to keep alive the tradition of the International Conference on Frontiers in Handwriting Recognition (ICFHR) that the TC-11 community decided to no longer organize as an independent conference during ICFHR 2022 held in Hyderabad, India. The handwriting track included oral sessions covering handwriting recognition for historical documents, synthesis of handwritten documents, as well as a subsection of one of the poster sessions. Additional presentation tracks at ICDAR 2023 featured Graphics Recognition, Natural Language Processing for Documents (D-NLP), Applications (including for medical, legal, and business documents), additional Document Analysis and Recognition topics (DAR), and a session highlighting featured competitions that were run for ICDAR 2023 (Competitions). Two poster presentation sessions were held at ICDAR 2023.

As ICDAR 2023 was held with in-person attendance, all papers were presented by their authors during the conference. Exceptions were only made for authors who could not attend the conference for unavoidable reasons. Such oral presentations were then provided by synchronous video presentations. Posters of authors that could not attend were presented by recorded teaser videos, in addition to the physical posters.

Three keynote talks were given by Marti Hearst (UC Berkeley), Vlad Morariu (Adobe Research), and Seichi Uchida (Kyushu University). We thank them for the valuable insights and inspiration that their talks provided for participants.

Finally, we would like to thank everyone who contributed to the preparation of the scientific program of ICDAR 2023, namely the authors of the scientific papers submitted to the journal track and directly to the conference, reviewers for journal-track papers, and both our regular and senior PC members. We also thank Ed Sobczak for helping with the conference web pages, and the ICDAR 2023 Publications Chairs Anurag Bharadwaj and Utkarsh Porwal, who oversaw the creation of this proceedings.

August 2023

Gernot A. Fink
Rajiv Jain
Koichi Kise
Richard Zanibbi

Organization

General Chairs

David Doermann   University at Buffalo, The State University of
                              New York, USA
Srirangaraj Setlur   University at Buffalo, The State University of
                              New York, USA

Executive Co-chairs

Venu Govindaraju   University at Buffalo, The State University of
                              New York, USA
Tong Sun   Adobe Research, USA

PC Chairs

Gernot A. Fink   Technische Universität Dortmund, Germany
                              (Europe)
Rajiv Jain   Adobe Research, USA (Industry)
Koichi Kise   Osaka Metropolitan University, Japan (Asia)
Richard Zanibbi   Rochester Institute of Technology, USA
                              (Americas)

Workshop Chairs

Mickael Coustaty   La Rochelle University, France
Alicia Fornes   Universitat Autònoma de Barcelona, Spain

Tutorial Chairs

Elisa Barney-Smith   Luleå University of Technology, Sweden
Laurence Likforman-Sulem   Télécom ParisTech, France

Competitions Chairs

Kenny Davila	Universidad Tecnológica Centroamericana, UNITEC, Honduras
Dimosthenis Karatzas	Universitat Autònoma de Barcelona, Spain
Chris Tensmeyer	Adobe Research, USA

Doctoral Consortium Chairs

Andreas Fischer	University of Applied Sciences and Arts Western Switzerland
Veronica Romero	University of Valencia, Spain

Publications Chairs

Anurag Bharadwaj	Northeastern University, USA
Utkarsh Porwal	Walmart, USA

Posters/Demo Chair

Palaiahnakote Shivakumara	University of Malaya, Malaysia

Awards Chair

Santanu Chaudhury	IIT Jodhpur, India

Sponsorship Chairs

Wael Abd-Almageed	Information Sciences Institute USC, USA
Cheng-Lin Liu	Chinese Academy of Sciences, China
Masaki Nakagawa	Tokyo University of Agriculture and Technology, Japan

Industry Chairs

Andreas Dengel	DFKI, Germany
Véronique Eglin	Institut National des Sciences Appliquées (INSA) de Lyon, France
Nandakishore Kambhatla	Adobe Research, India

Publicity Chairs

Sukalpa Chanda   Østfold University College, Norway
Simone Marinai   University of Florence, Italy
Safwan Wshah   University of Vermont, USA

Technical Chair

Edward Sobczak   University at Buffalo, The State University of New York, USA

Conference Secretariat

University at Buffalo, The State University of New York, USA

Program Committee

Senior Program Committee Members

Srirangaraj Setlur
Richard Zanibbi
Koichi Kise
Gernot Fink
David Doermann
Rajiv Jain
Rolf Ingold
Andreas Fischer
Marcus Liwicki
Seiichi Uchida
Daniel Lopresti
Josep Llados
Elisa Barney Smith
Umapada Pal
Alicia Fornes
Jean-Marc Ogier
C. V. Jawahar
Xiang Bai
Liangrui Peng
Jean-Christophe Burie
Andreas Dengel
Robert Sablatnig
Basilis Gatos

Apostolos Antonacopoulos
Lianwen Jin
Nicholas Howe
Marc-Peter Schambach
Marcal Rossinyol
Wataru Ohyama
Nicole Vincent
Faisal Shafait
Simone Marinai
Bertrand Couasnon
Masaki Nakagawa
Anurag Bhardwaj
Dimosthenis Karatzas
Masakazu Iwamura
Tong Sun
Laurence Likforman-Sulem
Michael Blumenstein
Cheng-Lin Liu
Luiz Oliveira
Robert Sabourin
R. Manmatha
Angelo Marcelli
Utkarsh Porwal

Program Committee Members

Harold Mouchere
Foteini Simistira Liwicki
Vernonique Eglin
Aurelie Lemaitre
Qiu-Feng Wang
Jorge Calvo-Zaragoza
Yuchen Zheng
Guangwei Zhang
Xu-Cheng Yin
Kengo Terasawa
Yasuhisa Fujii
Yu Zhou
Irina Rabaev
Anna Zhu
Soo-Hyung Kim
Liangcai Gao
Anders Hast
Minghui Liao
Guoqiang Zhong
Carlos Mello
Thierry Paquet
Mingkun Yang
Laurent Heutte
Antoine Doucet
Jean Hennebert
Cristina Carmona-Duarte
Fei Yin
Yue Lu
Maroua Mehri
Ryohei Tanaka
Adel M. M. Alimi
Heng Zhang
Gurpreet Lehal
Ergina Kavallieratou
Petra Gomez-Kramer
Anh Le Duc
Frederic Rayar
Muhammad Imran Malik
Vincent Christlein
Khurram Khurshid
Bart Lamiroy
Ernest Valveny
Antonio Parziale

Jean-Yves Ramel
Haikal El Abed
Alireza Alaei
Xiaoqing Lu
Sheng He
Abdel Belaid
Joan Puigcerver
Zhouhui Lian
Francesco Fontanella
Daniel Stoekl Ben Ezra
Byron Bezerra
Szilard Vajda
Irfan Ahmad
Imran Siddiqi
Nina S. T. Hirata
Momina Moetesum
Vassilis Katsouros
Fadoua Drira
Ekta Vats
Ruben Tolosana
Steven Simske
Christophe Rigaud
Claudio De Stefano
Henry A. Rowley
Pramod Kompalli
Siyang Qin
Alejandro Toselli
Slim Kanoun
Rafael Lins
Shinichiro Omachi
Kenny Davila
Qiang Huo
Da-Han Wang
Hung Tuan Nguyen
Ujjwal Bhattacharya
Jin Chen
Cuong Tuan Nguyen
Ruben Vera-Rodriguez
Yousri Kessentini
Salvatore Tabbone
Suresh Sundaram
Tonghua Su
Sukalpa Chanda

Mickael Coustaty
Donato Impedovo
Alceu Britto
Bidyut B. Chaudhuri
Swapan Kr. Parui
Eduardo Vellasques
Sounak Dey
Sheraz Ahmed
Julian Fierrez
Ioannis Pratikakis
Mehdi Hamdani
Florence Cloppet
Amina Serir
Mauricio Villegas
Joan Andreu Sanchez
Eric Anquetil
Majid Ziaratban
Baihua Xiao
Christopher Kermorvant
K. C. Santosh
Tomo Miyazaki
Florian Kleber
Carlos David Martinez Hinarejos
Muhammad Muzzamil Luqman
Badarinath T.
Christopher Tensmeyer
Musab Al-Ghadi
Ehtesham Hassan
Journet Nicholas
Romain Giot
Jonathan Fabrizio
Sriganesh Madhvanath
Volkmar Frinken
Akio Fujiyoshi
Srikar Appalaraju
Oriol Ramos-Terrades
Christian Viard-Gaudin
Chawki Djeddi
Nibal Nayef
Nam Ik Cho
Nicolas Sidere
Mohamed Cheriet
Mark Clement
Shivakumara Palaiahnakote
Shangxuan Tian

Ravi Kiran Sarvadevabhatla
Gaurav Harit
Iuliia Tkachenko
Christian Clausner
Vernonica Romero
Mathias Seuret
Vincent Poulain D'Andecy
Joseph Chazalon
Kaspar Riesen
Lambert Schomaker
Mounim El Yacoubi
Berrin Yanikoglu
Lluis Gomez
Brian Kenji Iwana
Ehsanollah Kabir
Najoua Essoukri Ben Amara
Volker Sorge
Clemens Neudecker
Praveen Krishnan
Abhisek Dey
Xiao Tu
Mohammad Tanvir Parvez
Sukhdeep Singh
Munish Kumar
Qi Zeng
Puneet Mathur
Clement Chatelain
Jihad El-Sana
Ayush Kumar Shah
Peter Staar
Stephen Rawls
David Etter
Ying Sheng
Jiuxiang Gu
Thomas Breuel
Antonio Jimeno
Karim Kalti
Enrique Vidal
Kazem Taghva
Evangelos Milios
Kaizhu Huang
Pierre Heroux
Guoxin Wang
Sandeep Tata
Youssouf Chherawala

Reeve Ingle
Aashi Jain
Carlos M. Travieso-Gonzales
Lesly Miculicich
Curtis Wigington
Andrea Gemelli
Martin Schall
Yanming Zhang
Dezhi Peng
Chongyu Liu
Huy Quang Ung
Marco Peer
Nam Tuan Ly
Jobin K. V.
Rina Buoy
Xiao-Hui Li
Maham Jahangir
Muhammad Naseer Bajwa

Oliver Tueselmann
Yang Xue
Kai Brandenbusch
Ajoy Mondal
Daichi Haraguchi
Junaid Younas
Ruddy Theodose
Rohit Saluja
Beat Wolf
Jean-Luc Bloechle
Anna Scius-Bertrand
Claudiu Musat
Linda Studer
Andrii Maksai
Oussama Zayene
Lars Voegtlin
Michael Jungo

Program Committee Subreviewers

Li Mingfeng
Houcemeddine Filali
Kai Hu
Yejing Xie
Tushar Karayil
Xu Chen
Benjamin Deguerre
Andrey Guzhov
Estanislau Lima
Hossein Naftchi
Giorgos Sfikas
Chandranath Adak
Yakn Li
Solenn Tual
Kai Labusch
Ahmed Cheikh Rouhou
Lingxiao Fei
Yunxue Shao
Yi Sun
Stephane Bres
Mohamed Mhiri
Zhengmi Tang
Fuxiang Yang
Saifullah Saifullah

Paolo Giglio
Wang Jiawei
Maksym Taranukhin
Menghan Wang
Nancy Girdhar
Xudong Xie
Ray Ding
Mélodie Boillet
Nabeel Khalid
Yan Shu
Moises Diaz
Biyi Fang
Adolfo Santoro
Glen Pouliquen
Ahmed Hamdi
Florian Kordon
Yan Zhang
Gerasimos Matidis
Khadiravana Belagavi
Xingbiao Zhao
Xiaotong Ji
Yan Zheng
M. Balakrishnan
Florian Kowarsch

Mohamed Ali Souibgui
Xuewen Wang
Djedjiga Belhadj
Omar Krichen
Agostino Accardo
Erika Griechisch
Vincenzo Gattulli
Thibault Lelore
Zacarias Curi
Xiaomeng Yang
Mariano Maisonnave
Xiaobo Jin
Corina Masanti
Panagiotis Kaddas
Karl Löwenmark
Jiahao Lv
Narayanan C. Krishnan
Simon Corbillé
Benjamin Fankhauser
Tiziana D'Alessandro
Francisco J. Castellanos
Souhail Bakkali
Caio Dias
Giuseppe De Gregorio
Hugo Romat
Alessandra Scotto di Freca
Christophe Gisler
Nicole Dalia Cilia
Aurélie Joseph
Gangyan Zeng
Elmokhtar Mohamed Moussa
Zhong Zhuoyao
Oluwatosin Adewumi
Sima Rezaei
Anuj Rai
Aristides Milios
Shreeganesh Ramanan
Wenbo Hu

Arthur Flor de Sousa Neto
Rayson Laroca
Sourour Ammar
Gianfranco Semeraro
Andre Hochuli
Saddok Kebairi
Shoma Iwai
Cleber Zanchettin
Ansgar Bernardi
Vivek Venugopal
Abderrhamne Rahiche
Wenwen Yu
Abhishek Baghel
Mathias Fuchs
Yael Iseli
Xiaowei Zhou
Yuan Panli
Minghui Xia
Zening Lin
Konstantinos Palaiologos
Loann Giovannangeli
Yuanyuan Ren
Shubhang Desai
Yann Soullard
Ling Fu
Juan Antonio Ramirez-Orta
Chixiang Ma
Truong Thanh-Nghia
Nathalie Girard
Kalyan Ram Ayyalasomayajula
Talles Viana
Francesco Castro
Anthony Gillioz
Huawen Shen
Sanket Biswas
Haisong Ding
Solène Tarride

Contents – Part VI

Posters: Scene Text

Text Reading Order in Uncontrolled Conditions by Sparse Graph
Segmentation 3
Renshen Wang, Yasuhisa Fujii, and Alessandro Bissacco

TDAE: Text Detection with Affinity Areas and Evolution Strategies 22
*Kefan Ma, Yuchen Luo, Zheng Huang, Kai Chen, Jie Guo,
and Weidong Qiu*

Visual Information Extraction in the Wild: Practical Dataset
and End-to-End Solution 36
*Jianfeng Kuang, Wei Hua, Dingkang Liang, Mingkun Yang,
Deqiang Jiang, Bo Ren, and Xiang Bai*

Scene Text Recognition with Image-Text Matching-Guided Dictionary 54
Jiajun Wei, Hongjian Zhan, Xiao Tu, Yue Lu, and Umapada Pal

E2TIMT: Efficient and Effective Modal Adapter for Text Image Machine
Translation 70
*Cong Ma, Yaping Zhang, Mei Tu, Yang Zhao, Yu Zhou,
and Chengqing Zong*

Open-Set Text Recognition via Shape-Awareness Visual Reconstruction 89
Chang Liu, Chun Yang, and Xu-Cheng Yin

Accelerating Transformer-Based Scene Text Detection and Recognition
via Token Pruning 106
Sergi Garcia-Bordils, Dimosthenis Karatzas, and Marçal Rusiñol

Text Enhancement: Scene Text Recognition in Hazy Weather 122
En Deng, Gang Zhou, Jiakun Tian, Yangxin Liu, and Zhenhong Jia

Reading Between the Lanes: Text VideoQA on the Road 137
*George Tom, Minesh Mathew, Sergi Garcia-Bordils,
Dimosthenis Karatzas, and C. V. Jawahar*

TPFNet: A Novel Text In-painting Transformer for Text Removal 155
Onkar Susladkar, Dhruv Makwana, Gayatri Deshmukh, Sparsh Mittal,
R. Sai Chandra Teja, and Rekha Singhal

Author Index 173

Posters: Scene Text

Text Reading Order in Uncontrolled Conditions by Sparse Graph Segmentation

Renshen Wang[(✉)], Yasuhisa Fujii, and Alessandro Bissacco

Google Research, Mountain View, USA
{rewang,yasuhisaf,bissacco}@google.com

Abstract. Text reading order is a crucial aspect in the output of an OCR engine, with a large impact on downstream tasks. Its difficulty lies in the large variation of domain specific layout structures, and is further exacerbated by real-world image degradations such as perspective distortions. We propose a lightweight, scalable and generalizable approach to identify text reading order with a multi-modal, multi-task graph convolutional network (GCN) running on a sparse layout based graph. Predictions from the model provide hints of bidimensional relations among text lines and layout region structures, upon which a post-processing cluster-and-sort algorithm generates an ordered sequence of all the text lines. The model is language-agnostic and runs effectively across multi-language datasets that contain various types of images taken in uncontrolled conditions, and it is small enough to be deployed on virtually any platform including mobile devices.

Keywords: Multi-modality · bidimensional ordering relations · graph convolutional networks

1 Introduction

Optical character recognition (OCR) technology has been developed to extract text reliably from various types of image sources [4]. Key components of an OCR system include text detection, recognition and layout analysis. As machine learning based digital image processing systems are nowadays ubiquitous and widely applied, OCR has become a crucial first step in the pipeline to provide text input for downstream tasks such as information extraction, text selection and screen reading.

Naturally, most image-to-text applications require very accurate OCR results to work well. This requirement is not only on text recognition—reading order among the recognized text lines is almost always as important as the recognition quality. The reason is self-evident for text selection (copy-paste) and text-to-speech tasks. And for structured document understanding like LayoutLM [33], DocFormer [3], FormNet [18], etc., the order of the input text also has a profound effect as most of these models have positional encoding attached to input text features, and a sequential labeling task for output. Input text order can sometimes be the key factor for the successful extraction of certain entities.

© The Author(s), under exclusive license to Springer Nature Switzerland AG 2023
G. A. Fink et al. (Eds.): ICDAR 2023, LNCS 14192, pp. 3–21, 2023.
https://doi.org/10.1007/978-3-031-41731-3_1

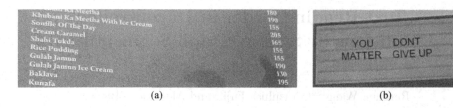

Fig. 1. Hard examples for text reading order. (a) A cropped image of a menu with dish names and prices, where a correct reading order necessarily needs correct association between each dish name and its price, which is a hard task for humans without the full image context due to the perspective distortion in the image. (b) A text layout intentionally made to have two different reading order interpretations, both valid, but with completely opposite meanings.

Depending on the text layout, the difficulty of deciding its reading order varies greatly. It can be as simple as sorting all the text lines by y-coordinates, but can also be hard like the images in Fig. 1. Even if we exclude corner cases like these, there are still complexities brought by the diversity of layout structures which are often domain specific. Previous studies have tackled the problem in different ways. Rule based approaches like [1, 9, 27] usually aim at one specific domain, while learning based approaches like [6, 21, 32] are more general but have scalability issues (more discussions in the following section).

In this paper, we propose a composite method that uses both machine learning model and rule based sorting to achieve best results. It is based on the observation from [1] that most reading order sequences are in one of the two patterns—column-wise and row-wise—as illustrated in Fig. 2.

We use a graph convolutional network that takes spatial-image features from the input layout and image, and segments the layout into two types of regions where the paragraphs can be properly sorted by the type of their patterns. A β-skeleton graph built on boxes [31] enables efficient graph convolutions while also providing edge bounding boxes for RoI (regions of interest) pooling from the image feature map. A post-processing cluster-and-sort algorithm finalizes the overall reading order based on model predictions. This unique combination gives us an effective, lightweight, scalable and generalizable reading order solution.

2   Related Work

Two types of related work are discussed in this section. The first subsection includes previous reading order efforts, and the second subsection discusses other multi-modal image-text-spatial models that share some of the components with our approach.

2.1   Reading Order Detection

Previous studies have tackled the reading order problem in various ways. We roughly categorize them into rule based sorting [1, 5, 9, 27] and machine-learning based sequence prediction [6, 21, 29, 32], etc.

(a)   (b)

Fig. 2. Two major patterns of reading order. (a) Column-wise order, most common in printed media like newspapers and magazines. (b) Row-wise order, usually in receipts, forms and tabular text blocks.

Topological sort was proposed in [5] for document layout analysis where partial orders are based on x/y interval overlaps among text lines. It can produce reading order patterns like Fig. 2(a) for multi-column text layouts. A bidimensional relation rule proposed in [1] provides similar topological rules, and in addition provides a row-wise rule by inverting the x/y axes from column-wise. An argumentation based approach in [9] works on similar rules derived from text block relations. For large text layout with hierarchies, XY-Cut [13, 27] can be an effective way for some layout types to order all the text blocks top-to-bottom and left-to-right. These rule based approaches can work accurately for documents in certain domains. But without extra signals, they will fail for out-of-domain cases like Fig. 2(b).

Machine learning based approaches are designed to learn from training examples across different domains to enable a general solution. The data mining approach in [6] learns partial order among text blocks from their spatial features and identifies reading order chains from the partial orders. A similar approach in [29] trains a model to predict pairwise order relations among text regions and curves for handwritten documents. The major limitation is that any partial order between two entities are derived from their own spatial features without the layout structure information in their neighborhood. So these models may not be able to identify the layout structure among a group of text lines and therefore fail to find the correct pattern.

Graph convolutional networks and transformer models provide mechanisms for layout-aware signals by interactions between layout entities. A text reorganization model introduced in [21] uses a graph convolutional encoder and a pointer network decoder to reorder text blocks. With a fully-connected graph at its input, the graph encoder functions similarly as a transformer encoder. Image features are added to graph nodes by RoI pooling on node boxes with bilinear interpolation. Another work LayoutReader [32] uses a transformer based architecture on spatial-text features instead of spatial-image features to predict reading order sequence on words. The text features enable it to use the powerful LayoutLM [34] model, but also make it less generalizable. These models are capable of predicting reading order within complex layout structures. However, there are scalability issues in two aspects:

– Run time scales quadratically with input size. Whether in the graph convolutional encoder with full connections or the sequence pointer decoder, most of the components have $O(n^2)$ time complexity, and may become too slow for applications with dense text.
– Accuracy scales inversely with input size. The fully-connected self-attention mechanism in the encoder takes all the text entities to calculate a global attention map, which introduces noises to the reading order signals that should be decidable from local layout structures. The sequence decoder uses softmax probabilities to determine the output index for each step, where the output range increases with input size, and so does the chance of errors. Figure 10 illustrates this limitation from our experiments.

To summarize briefly, there are multiple effective ways to order OCR text by rule based or machine learning based methods, and in both categories there is room for improvement in generalizability and scalability.

2.2   Spatial, Image Features and Multi-Modality

Multi-modal transformer models have become mainstream for document or image understanding tasks. Related work include LayoutLM [13,15,33,34], Doc-Former [3], SelfDoc [22], UDoc [12], StrucText [23], TILT [28], LiLT [30], Form-Net [18], PaLI [7], etc.

Document image understanding starts with an OCR engine that provides text content as the main input for the language model. Alongside, the text bounding boxes associated with the words and lines provide important spatial features (sometimes called layout features or geometric features). Additionally, since not all visual signals are captured by the OCR engine, an image component in the model can help cover the extra contextual information from the input. Thus, a model to achieve best results should take all of the three available modalities.

For image features, most previous studies use RoI pooling [8] by the text bounding boxes from OCR, and the pooled features are attached to the corresponding text entity. It is effective for capturing text styles or colors, but less so for visual cues out of those bounding boxes, such as the curly separation lines in Fig. 3. While it is possible to use an image backbone with large receptive fields, like ResNet50 used in the UDoc model or U-Net used in the TILT model, it is not an ideal solution for two reasons:

– In sparse documents, useful visual cues can be far from any text on the page.
– Large receptive fields bring in extra noise from regions irrelevant to the features we need.

Thus, it will be more effective to have image RoI boxes that cover pairs of text bounding boxes. A sparse graph like β-skeleton used in [31] can provide the node pairs for such RoI pooling without significantly increasing the model's memory footprint and computational cost.

3   Proposed Method

Based on previous studies, we design a lightweight machine learning based app-roach with a model that is small in size, fast to run, and easy to generalize in uncontrolled conditions.

3.1   Strong Patterns of Reading Order

From a set of real-world images annotated with reading order, we have an obser-vation that matches very well with the bidimensional document encoding rules in [1]—column-wise text usually has a zigzag pattern of Fig. 2(a), and row-wise text has a similar but transposed zigzag like Fig. 2(b). Some images may con-tain both types of text, which makes the pattern more complex. But once the column-wise/row-wise type of a text region is decided, the reading order in this region mostly follows the pattern and can be determined with a topological sort according to the bidimensional rules. Figure 7(a) shows an example of an annotated reading order sequence.

Based on this observation, learning text reading order becomes an image segmentation problem, as opposed to learning arbitrary global sequences of text entities. Instead of predicting the next entity in the entire image, we do a binary classification for each text entity on whether it's in a column-wise or row-wise pattern. Moreover, the pattern classification for a text line can be decided by local layout structures, and global attention maps are therefore unnecessary.

3.2   Model Architecture

We use a graph convolutional network (GCN) with a sparse graph construction because of the three major advantages listed here:

- GCN models are equivariant to input order permutations. It is natural to assume that a model deciding reading order should not depend on the order of its input.
- With a sparse graph like β-skeleton, GCN computation scales linearly with input size.
- Graph edges constructed from text boxes can provide edge bounding boxes, which are better for image feature RoI pooling (Fig. 3, Table 2).

As illustrated in Fig. 4, we use an MPNN [11] variant of GCN as the main model backbone, and a β-skeleton graph [17] constructed with text line boxes as nodes. Similar configurations have been applied to other layout problems [18, 19, 25, 31], and graph construction details are available in [31]. The main GCN input is from the spatial features of text line bounding boxes as node features, including x, y coordinate values of the box corners, and the coordinate values multiplied by rotation angle coefficients $\cos \alpha$, $\sin \alpha$. The spatial features go through T steps of graph convolution layers, each containing a node-to-edge "message passing" layer and edge-to-node aggregation layer with attention weighted pooling.

Fig. 3. A cropped example of a β-skeleton graph [31] constructed from text line boxes. Graph node boxes are shown in green and edge lines in cyan. The three orange colored text lines demonstrate how image features can help—the 2nd and 3rd boxes are closer in distance, so spatial features may indicate they are in the same section, but the curly separation line between them indicates otherwise. The yellow box at the bottom is the minimum containing box of the two line boxes inside, where the RoI pooling can cover image features between these lines. (Color figure online)

Besides the main input from nodes, we add a side input of edge features from edge box RoI pooling on an image feature map to help capture potential visual cues surrounding text boxes. We use MobileNetV3-Small [14] as the image backbone for its efficiency. Note that the purpose of this image backbone is not for a major task like object detection, but to look for auxiliary features like separation lines and color changes, so a small backbone is capable enough for our task. For the same reason, we reduce the MobileNetV3 input image size to 512×512 to speed up training and inference. The details of the image processing are illustrated in Fig. 5. In most cases, the text content is no longer recognizable

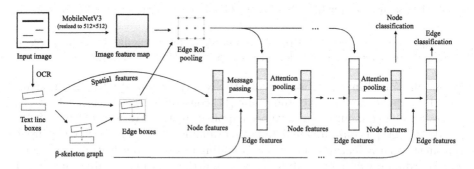

Fig. 4. Overview of the reading order multi-classifier model. Node classification predicts the reading order patterns, and edge classification predicts paragraph clustering.

Fig. 5. Image processing for the MobileNetV3 input. The inner yellow box is the minimum containing box of all the text lines in the image. If its diagonal d is larger than 512, we scale down the image by $\frac{512}{d}$ so that all the text bounding boxes are contained in the white circle of diameter 512, and then we crop (maybe also pad) around this circle to get the final processed image. This process ensures that both the image and the layout can be randomly rotated during training without any line box moved out of boundary. (Color figure online)

after such downsizing, but the auxiliary image features can be well preserved. We also make sure that the entire layout is contained in a circle of diameter 512 within the processed image, which enables random rotations during model training—a key augmentation for our model to work in all conditions.

Language features are not included in order to keep the model minimal in size and independent of domain knowledge. Also, our annotated reading order data is limited in English only, upon which we try to train a universal model.

The GCN is a multi-task model that outputs both node and edge predictions. At node level, it predicts the reading order pattern on each line box (column-wise or row-wise). These predictions are essentially a segmentation for text regions where the lines can be sorted accordingly.

At edge level, the model predicts whether the two lines connected by an edge belong to the same paragraph. Thus, it works like the edge clustering models in [25,31], and we can improve the final reading order by grouping lines together within each paragraph. The reading order estimation by the grouping usually do not affect column-wise order among text lines, but can be critical in row-wise regions such as tables or forms with multi-line cells, e.g. Fig. 9(d).

It may be considered that a fully convolutional network can do similar segmentation tasks like [16,26] on the input image. However, we have observed that such models are less effective for certain types of text content—e.g. in Fig. 2(b), similar lines in the left column are grouped into a large paragraph, disrupting the row-wise reading order.

3.3 Recovering Reading Order from Model Predictions

With the β-skeleton graph that provides local connections among dense text boxes, the GCN model predicts on *local* properties of the text, which can be aggregated to give us a *global* reading order. To handle mixed column-wise and row-wise predictions as well as potential text rotations and distortions in the input image, we extend the rule based sorting in [1,5] and propose a hierarchical cluster-and-sort algorithm to recover the global reading order from line-level pattern predictions and clustered paragraphs. The following Algorithm 1 generates a set of clusters, each cluster c_i contains a non-empty set of paragraphs and maybe a set of child clusters. Each cluster is also assigned a reading order pattern $R(c_i) \in \{col, row\}$, with *col* for column-wise and *row* for row-wise.

Row-wise text often involves sparse tables with components not directly connected by β-skeleton edges, so the hop edges like in [25] can be helpful in step 4 of algorithm 1. More details can be added, e.g. setting an edge length threshold in step 3 to avoid merging distant clusters.

Algorithm 1: Hierarchical Clustering

Input: Text line bounding boxes, β-skeleton graph G,
   GCN node predictions and edge predictions.

---

1. Cluster lines into paragraphs $p_1, ..., p_n$ from edge predictions.
2. Each paragraph is initialized as a cluster, $c_i = \{p_i\}$. Reading order pattern $R(c_i)$ is the majority vote from the paragraph's line predictions .
3. For each edge $(i, j) \in G$, find cluster c_a containing line i and c_b containing line j; if $R(c_a) = R(c_b) = col$, merge c_a and c_b into a bigger column-wise cluster.
4. For each edge $(i, j) \in G$ or hop edge (i, j) ($\exists k$ that $(i, k) \in G$ and $(k, j) \in G$), find cluster c_a containing line i and c_b containing line j; if $R(c_a) = R(c_b) = row$, merge c_a and c_b into a bigger row-wise cluster.
5. Calculate the containing box for each cluster. The rotation angle of the box is the circular mean angle of all the paragraphs in the cluster.
6. Sort the clusters by ascending area of their containing boxes.
7. For each cluster c_i, if its containing box $B(c_i)$ overlaps with $B(c_j)$ by area greater than $T \times Area(B(c_i))$, set c_i as a child cluster of c_j.
8. Create a top level cluster with all the remaining clusters as its children.

Once the regions of reading order patterns are decided by the hierarchical clusters, we can use topological sort within each cluster as in Algorithm 2.

With all the clusters sorted, an ordered traversal of the cluster hierarchy can give us the final reading order among all the paragraphs. Figure 6 shows the reading order on a packaging box at different camera angles. Note that the algorithms are not sensitive to bounding box angles, and the model is trained with randomly augmented data, so the rotation has minimal effect on the final result. It can even handle vertical text lines in Chinese/Japanese with the vertical lines regarded as rotated horizontal lines.

Algorithm 2: Reading Order Sorting within a Cluster

Input: Bounding boxes $b_1, ..., b_n$ from paragraphs or child
clusters, the reading order pattern to sort with.

---

1. Calculate α, the circular mean angle from all the bounding box angles.
2. For each box b_i, rotate it around $(0,0)$ by angle $-\alpha$.
3. For each box b_i, calculate its axis aligned minimum containing box a_i.
4. If the reading order pattern is column-wise,
   Add constraint $(i \rightarrow j)$ if a_i, a_j overlap on x-axis and
   $$y_{center}(a_i) < y_{center}(a_j)$$
   Sort $a_1, ..., a_n$ by ascending x_{center}
   else   # pattern is row-wise
   Add constraint $(i \rightarrow j)$ if a_i, a_j overlap on y-axis and
   $$x_{center}(a_i) < x_{center}(a_j)$$
   Sort $a_1, ..., a_n$ by ascending y_{center}
5. Based on existing order, topologically sort $a_1, ..., a_n$ with the order constraints.

3.4 Data Labeling

We prepared a dataset with human annotated layout data, including paragraphs
as polygons and reading order groups where each group is an ordered sequence
of paragraphs. Figure 7(a) shows a set of paragraphs, where the reading order
starts with the green paragraph and follows the jagged line.

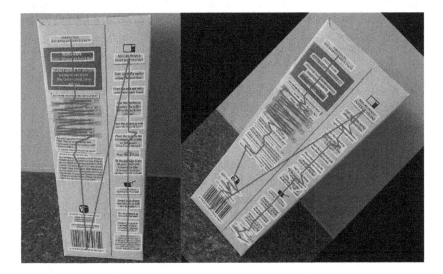

Fig. 6. Reading order example at different angles. Paragraphs with column-wise pattern predictions are shown in yellow, row-wise in pink. The dark blue line shows the overall reading order among all paragraphs. (Color figure online)

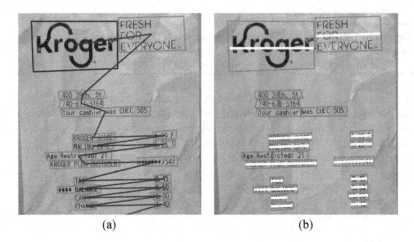

(a)   (b)

Fig. 7. Labeling reading order patterns from annotations. (a) Ground truth from human annotated paragraphs and reading order. (b) Reading order pattern inferred from the annotated sequence—column-wise indicated by a vertical/purple line in each paragraph and row-wise by a horizontal/green line. (Color figure online)

Algorithm 3: Pattern Labeling from Annotated Reading Order

Input: A sequence of ground truth paragraphs $p_1, p_2, p_3, \cdots, p_n$ represented as rectangular boxes.

---

1. Between each consecutive pair of paragraphs (p_i, p_{i+1}), we categorize their geometrical relation $R_{i,i+1}$ as one of $\{vertical, horizontal, unknown\}$.

   (a) Calculate α, the circular mean angle of the two boxes' rotation angles.

   (b) Rotate the boxes of p_i and p_{i+1} around $(0, 0)$ by $-\alpha$, denoted as b_i and b_{i+1}.

   (c) Axis aligned box c is the minimum containing box of both b_i and b_{i+1}.

   (d) if $y_{overlap}(b_i, b_{i+1}) < 0.1 \cdot height(c)$ and $y_{center}(b_i) < y_{center}(b_{i+1})$
   $$R_{i,i+1} = vertical$$

   (e) else if $x_{overlap}(b_i, b_{i+1}) < 0.1 \cdot width(c)$ and $x_{center}(b_i) < x_{center}(b_{i+1})$
   if c does not cover paragraphs other than p_i, p_{i+1}
   $$R_{i,i+1} = horizontal \qquad \text{# mostly tabular structures}$$
   else
   $$R_{i,i+1} = vertical \qquad \text{# mostly multi-column text}$$

   (f) In other conditions, $R_{i,i+1} = unknown$

2. Decide the reading order pattern for paragraph p_i from $R_{i-1,i}$ and $R_{i,i+1}$.

   (a) $(unknown, unknown) \rightarrow unknown$

   (b) In case of one unknown, the other one decides the pattern: $vertical \rightarrow$ column-wise, $horizontal \rightarrow$ row-wise,.

   (c) If neither is unknown, $(vertical, vertical) \rightarrow$ column-wise, otherwise it is row-wise.

While the edge clustering labels are straightforward from the paragraph polygons, the reading order pattern labeling is less trivial because we need to derive binary labels from ground truths of paragraph ordering. We decide the pattern of a paragraph by comparing its position with its predecessor and successor. Figure 7(b) shows an example, and detailed logic is elaborated in Algorithm 3.

3.5 Limitations

The node-edge classification model can produce reasonable reading order in most cases, but may fail for complex layouts with multiple tabular sections placed closely, like the cross section errors in Fig. 11(a). The root cause is the lack of higher level layout structure parsing with the two classification tasks. Data annotation at section level is generally hard because there is no universal agreement on the exact definition of sections among text. Figure 11(b) shows the result with extra section level clustering trained on a domain specific dataset. There is significant improvement, yet cross domain generalization is not guaranteed, and we can still see imperfections in the multi-section reading order due to section prediction errors.

Fig. 8. Cluster-and-sort result on the cropped menu from Fig. 1(a). Although the model correctly predicts the row-wise pattern, reading order is still incorrect due to the perspective distortion and the unusually large spacing between the two columns.

Another limitation is that our model is not a reliable source for parsing table structures like [24]. Figure 8 shows the reading order result of the image in Fig. 1(a). Note that in the sorting algorithm, we rotate all the bounding boxes to zero out their mean angle. But when the boxes are at different angles due to distortions, there will still be slanted line boxes and misaligned table rows after all the rotations, so the topological sort on the axis-aligned containing boxes cannot guarantee the right order. In presence of tables, a separate model with structure predictions will likely perform better.

4   Experiments

We experiment with the GCN model with predictions on reading order pattern and paragraph clustering, together with the cluster-and-sort algorithms.

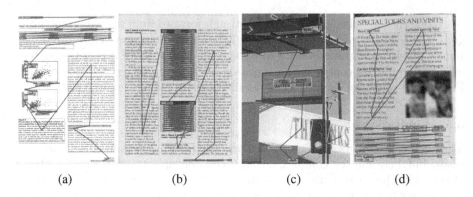

(a)	(b)	(c)	(d)

Fig. 9. Reading order results from (a) PubLayNet [35], (b) PRIMA Contemporary dataset [2], (c) the ambiguous example from Fig. 1 with a positive interpretation, and (d) our evaluation set.

4.1 Datasets and Evaluation Metrics

Various metrics have been used to evaluate reading order, such as Spearman's footrule distance, Kendall's Tau rank distance used in [29] and BLEU scores in [21]. These metrics can accurately measure order mismatches, but also require full length ground truth order for comparison.

We created an annotated layout dataset where reading order ground truths are partially annotated, i.e. some subsets of paragraphs form reading order groups with annotated order, and the order among groups is undefined. This makes it more flexible to match realistic user requirements and less suitable for full ranking metrics. So instead, we use a normalized Levenshtein distance [20] which measures the minimum number of word operations (insertions and deletions) needed to equalize two lists. For each reading order group, we take the ordered list of paragraphs and find all the OCR words W contained in these polygons. The word order within each paragraph is taken directly from OCR (mostly accurate for a single paragraph). Then we find the shortest subsequence of the serialized OCR output that contains all the words in W, compute its Levenshtein distance to W, and multiply it by the normalization factor $1/|W|$.

Besides our annotated set, we test the model with PubLayNet [35] because of its variety on layout components with different reading order patterns. Although there is no ground truth of reading order, we take "text" instances as paragraphs with column-wise pattern, and "table"/"figure" types as containers of text lines with row-wise pattern. Thus, we are able to train the same multi-task GCN model. The annotated set contains 25K text images in English for training and a few hundred test images for each of the available languages, and PubLayNet contains 340K training images and 12K validation images all in English.

4.2 Model Setup

The model is built as shown in Fig. 4, with the OCR engine from Google Cloud Vision API producing text lines and their spatial features. Edge image features are from a bi-linear interpolation on the MobileNetV3 output with 16×3 points each box and dropout rate 0.5. The TF-GNN [10] based GCN backbone uses 10 steps of weight-sharing graph convolutions, with node feature dimension 32 and message passing hidden dimension 128. Edge-to-node pooling uses a 4-head attention with 3 hidden layers of size 16 and dropout rate 0.5. Total number of parameters is 267K including 144K from MobileNetV3-Small.

We train the model for 10M steps with randomized augmentations including rotation and scaling, so the model can adapt to a full range of inputs. The OCR boxes are transformed together with the image in each training example, resulting in better robustness than previous approaches (Fig. 6).

4.3   Baselines

Most commercial OCR systems use a topological sort like in [1] with one of the two patterns. We use column-wise pattern in the basic baseline as it produces better scores than row-wise in our evaluations, and is close to the default output order from the OCR engine we use.

In addition, we implement a GCN model that directly predicts edge directions on a fully connected graph similar to the model in [21]. Figure 10 shows two examples with comparison between this baseline and our approach, with supports the scalability discussion in Subsect. 2.1.

4.4   Results

We train the multi-task model with PubLayNet and our paragraph reading order set added with the menu photos labelled from human annotations. From Table 1, we can see the difference in the difficulty between the two sets. Real-world images from our dataset have much larger variations on layout styles and image degradations that make the same tasks much harder to learn.

Table 1. Scores of the two classification tasks on PubLayNet and our labelled paragraph reading order dataset.

Dataset	Reading order pattern			Paragraph clustering		
	Precision	Recall	F1	Precision	Recall	F1
PubLayNet	0.998	0.995	0.997	0.994	0.996	0.995
Annotated ordered paragraphs	0.828	0.805	0.819	0.895	0.909	0.902

Table 2. F1 scores from the image feature ablation test.

Boxes for image feature RoI pooling	Reading order pattern			Paragraph clustering		
	Precision	Recall	F1	Precision	Recall	F1
n/a	0.800	0.803	0.802	0.887	0.895	0.891
Node boxes	0.819	0.781	0.800	0.870	0.903	0.886
Edge boxes	0.828	0.805	0.819	0.895	0.909	0.902

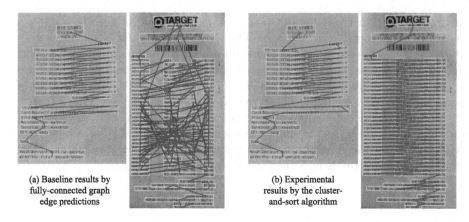

(a) Baseline results by fully-connected graph edge predictions

(b) Experimental results by the cluster-and-sort algorithm

Fig. 10. Comparison between the fully-connected graph model and our approach on two receipt examples. The full graph predictions perform well on the sparse example, but fail on the dense one.

We also test the effectiveness of the edge box RoI pooling by an image feature ablation test, where the baseline is the model with all image features removed, compared against ones with node box RoI pooling and edge box RoI pooling. Table 2 shows that node box RoI does not help at all, even with a slight accuracy drop compared with the baseline. These results confirm our previous hypothesis that the image backbone mainly helps the model by discovering visual cues out of text bounding boxes, and edge boxes are much more effective for this purpose.

Finally, we measure the normalized Levenshtein distance for reading order produced by the GCN and the cluster-and-sort algorithm, and compare it against the two baseline methods in Subsect. 4.3. As in Table 3, our algorithm can greatly improve reading order quality across all Latin languages, even though the training data is only available in English. The model also works well for examples out

Table 3. Normalized Levenshtein distance (lower is better) on a multi-language reading order evaluation set. Training data is only available in English.

Language	Training set size	Test set size	All-column-wise baseline	Fully-connected graph baseline	2-task GCN cluster-and-sort
English	25K	261	0.146	0.126	0.098
French	n/a	218	0.184	0.144	0.119
Italian		189	0.172	0.145	0.122
German		196	0.186	0.162	0.112
Spanish		200	0.183	0.103	0.097
Russian		1003	0.202	0.159	0.148
Hindi		990	0.221	0.181	0.152
Thai		951	0.131	0.111	0.104

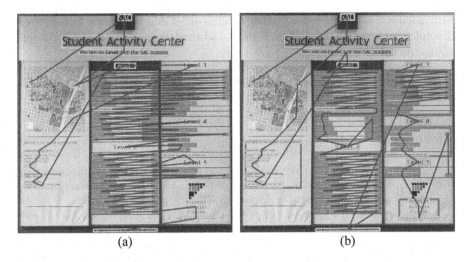

(a)   (b)

Fig. 11. A multi-section example. (a) Paragraphs with row-wise pattern are clustered into overly large regions, causing incorrect cross-section reading order. (b) With section level clusters (shown in orange) added into Algorithm 1, multi-table results can be improved. (Color figure online)

of our datasets. Figure 9 includes images from various sources, demonstrating the effectiveness of our model with inputs ranging from digital/scanned documents to scene images.

5   Conclusions and Future Work

We show that GCN is highly efficient at predicting reading order patterns and various layout segmentation tasks, which is further enhanced with a small image backbone providing edge RoI pooled signals. Our model is small in size and generalizes well enough to be deployable on any platform to improve OCR quality or downstream applications.

In addition, the GCN model has the potential to handle more than two tasks. We tried an extra edge prediction task trained with a dataset of menu photos with section level polygon annotations. Unlike general document or scene text images, menus like Fig. 3 usually have clearly defined sections like main dishes, side dishes, drinks, etc. Therefore, the menu dataset has accurate and consistent section level ground truth for model training. The 3-task GCN model provides higher-level layout information to the clustering algorithm and helps produce Fig. 11(b), a major improvement on reading order. Still, there is domain specific knowledge on menu sections that does not always generalize well. And because most evaluation examples have relatively simple layouts, the 3-task model has not produced better results than the 2-task model in our experiments. Nevertheless, we think section level ground truth or higher-level layout structural information will be valuable for further reading order improvements. Future work will explore the possibilities of both data and modeling approaches for parsing layout structures.

Acknowledgements. The authors would like to thank Ashok C. Popat and Chen-Yu Lee for their valuable reviews and feedback.

References

1. Aiello, M., Smeulders, A.M.W.: Bidimensional relations for reading order detection. In: EPRINTS-BOOK-TITLE. University of Groningen, Johann Bernoulli Institute for Mathematics and Computer Science (2003)
2. Antonacopoulos, A., Bridson, D., Papadopoulos, C., Pletschacher, S.: A realistic dataset for performance evaluation of document layout analysis. In: 10th International Conference on Document Analysis and Recognition, ICDAR 2009, Barcelona, Spain, 26–29 July 2009, pp. 296–300. IEEE Computer Society (2009). https://doi.org/10.1109/ICDAR.2009.271
3. Appalaraju, S., Jasani, B., Kota, B.U., Xie, Y., Manmatha, R.: Docformer: end-to-end transformer for document understanding. In: 2021 IEEE/CVF International Conference on Computer Vision, ICCV 2021, Montreal, QC, Canada, 10–17 October 2021, pp. 973–983. IEEE (2021). https://doi.org/10.1109/ICCV48922.2021.00103
4. Bissacco, A., Cummins, M., Netzer, Y., Neven, H.: Photoocr: reading text in uncontrolled conditions. In: IEEE International Conference on Computer Vision, ICCV 2013, Sydney, Australia, 1–8 December 2013, pp. 785–792. IEEE Computer Society (2013). https://doi.org/10.1109/ICCV.2013.102
5. Breuel, T.M.: High performance document layout analysis. In: Symposium on Document Image Understanding Technology, Greenbelt, MD, USA (2003)
6. Ceci, M., Berardi, M., Porcelli, G., Malerba, D.: A data mining approach to reading order detection. In: 9th International Conference on Document Analysis and Recognition (ICDAR 2007), 23–26 September, Curitiba, Paraná, Brazil, pp. 924–928. IEEE Computer Society (2007). https://doi.org/10.1109/ICDAR.2007.4377050
7. Chen, X., et al.: Pali: a jointly-scaled multilingual language-image model (2022). https://doi.org/10.48550/ARXIV.2209.06794. https://arxiv.org/abs/2209.06794
8. Dai, J., Li, Y., He, K., Sun, J.: R-FCN: object detection via region-based fully convolutional networks. In: Lee, D.D., Sugiyama, M., von Luxburg, U., Guyon, I., Garnett, R. (eds.) Advances in Neural Information Processing Systems 29: Annual Conference on Neural Information Processing Systems 2016, 5–10 December 2016, Barcelona, Spain, pp. 379–387 (2016)
9. Ferilli, S., Grieco, D., Redavid, D., Esposito, F.: Abstract argumentation for reading order detection. In: Simske, S.J., Rönnau, S. (eds.) ACM Symposium on Document Engineering 2014, DocEng 2014, Fort Collins, CO, USA, 16–19 September 2014, pp. 45–48. ACM (2014). https://doi.org/10.1145/2644866.2644883
10. Ferludin, O., et al.: TF-GNN: graph neural networks in tensorflow (2022). https://doi.org/10.48550/ARXIV.2207.03522. https://arxiv.org/abs/2207.03522
11. Gilmer, J., Schoenholz, S.S., Riley, P.F., Vinyals, O., Dahl, G.E.: Neural message passing for quantum chemistry. In: Proceedings of the 34th International Conference on Machine Learning, ICML 2017, vol. 70, pp. 1263–1272. JMLR.org (2017)

12. Gu, J., et al.: Unidoc: unified pretraining framework for document understanding. In: Ranzato, M., Beygelzimer, A., Dauphin, Y.N., Liang, P., Vaughan, J.W. (eds.) Advances in Neural Information Processing Systems 34: Annual Conference on Neural Information Processing Systems 2021, NeurIPS 2021, 6–14 December 2021, Virtual, pp. 39–50 (2021)

13. Gu, Z., et al.: Xylayoutlm: towards layout-aware multimodal networks for visually-rich document understanding. In: Proceedings of the IEEE/CVF Conference on Computer Vision and Pattern Recognition (CVPR), pp. 4583–4592 (2022)

14. Howard, A., et al.: Searching for mobilenetv3. In: 2019 IEEE/CVF International Conference on Computer Vision, ICCV 2019, Seoul, Korea (South), 27 October–2 November 2019, pp. 1314–1324. IEEE (2019). https://doi.org/10.1109/ICCV.2019.00140

15. Huang, Y., Lv, T., Cui, L., Lu, Y., Wei, F.: Layoutlmv3: pre-training for document AI with unified text and image masking. CoRR **abs/2204.08387** (2022). https://doi.org/10.48550/ARXIV.2204.08387. https://arxiv.org/abs/2204.08387

16. Kirillov, A., He, K., Girshick, R.B., Rother, C., Dollár, P.: Panoptic segmentation. In: IEEE Conference on Computer Vision and Pattern Recognition, CVPR 2019, Long Beach, CA, USA, 16–20 June 2019, pp. 9404–9413. Computer Vision Foundation/IEEE (2019). https://doi.org/10.1109/CVPR.2019.00963

17. Kirkpatrick, D.G., Radke, J.D.: A framework for computational morphology. Mach. Intell. Pattern Recognit. **2**, 217–248 (1985). https://doi.org/10.1016/B978-0-444-87806-9.50013-X

18. Lee, C., et al.: Formnet: structural encoding beyond sequential modeling in form document information extraction. In: Muresan, S., Nakov, P., Villavicencio, A. (eds.) Proceedings of the 60th Annual Meeting of the Association for Computational Linguistics (Volume 1: Long Papers), ACL 2022, Dublin, Ireland, 22–27 May 2022, pp. 3735–3754. Association for Computational Linguistics (2022). https://aclanthology.org/2022.acl-long.260

19. Lee, C., et al.: ROPE: reading order equivariant positional encoding for graph-based document information extraction. In: Zong, C., Xia, F., Li, W., Navigli, R. (eds.) Proceedings of the 59th Annual Meeting of the Association for Computational Linguistics and the 11th International Joint Conference on Natural Language Processing, ACL/IJCNLP 2021, (Volume 2: Short Papers), Virtual Event, 1–6 August 2021, pp. 314–321. Association for Computational Linguistics (2021). https://doi.org/10.18653/v1/2021.acl-short.41

20. Levenshtein, V.I.: Binary codes capable of correcting deletions, insertions and reversals. Soviet Physics Doklady **10**, 707 (1966)

21. Li, L., Gao, F., Bu, J., Wang, Y., Yu, Z., Zheng, Q.: An end-to-end OCR text re-organization sequence learning for rich-text detail image comprehension. In: Vedaldi, A., Bischof, H., Brox, T., Frahm, J.-M. (eds.) ECCV 2020. LNCS, vol. 12370, pp. 85–100. Springer, Cham (2020). https://doi.org/10.1007/978-3-030-58595-2_6

22. Li, P., et al.: Selfdoc: self-supervised document representation learning. In: IEEE Conference on Computer Vision and Pattern Recognition, CVPR 2021, virtual, 19–25 June 2021, pp. 5652–5660. Computer Vision Foundation/IEEE (2021)

23. Li, Y., et al.: Structext: structured text understanding with multi-modal transformers. In: Shen, H.T., et al. (eds.) MM 2021: ACM Multimedia Conference, Virtual Event, China, 20–24 October 2021, pp. 1912–1920. ACM (2021). https://doi.org/10.1145/3474085.3475345

24. Liu, H., Li, X., Liu, B., Jiang, D., Liu, Y., Ren, B.: Neural collaborative graph machines for table structure recognition. In: Proceedings of the IEEE/CVF Conference on Computer Vision and Pattern Recognition (CVPR), pp. 4533–4542 (2022)
25. Liu, S., Wang, R., Raptis, M., Fujii, Y.: Unified line and paragraph detection by graph convolutional networks. In: Uchida, S., Barney, E., Eglin, V. (eds.) DAS 2022. LNCS, vol. 13237, pp. 33–47. Springer, Cham (2022). https://doi.org/10.1007/978-3-031-06555-2_3
26. Long, J., Shelhamer, E., Darrell, T.: Fully convolutional networks for semantic segmentation. In: IEEE Conference on Computer Vision and Pattern Recognition, CVPR 2015, Boston, MA, USA, 7–12 June 2015, pp. 3431–3440. IEEE Computer Society (2015). https://doi.org/10.1109/CVPR.2015.7298965
27. Meunier, J.: Optimized XY-cut for determining a page reading order. In: Eighth International Conference on Document Analysis and Recognition (ICDAR 2005), 29 August–1 September 2005, Seoul, Korea, pp. 347–351. IEEE Computer Society (2005). https://doi.org/10.1109/ICDAR.2005.182
28. Powalski, R., Borchmann, Ł., Jurkiewicz, D., Dwojak, T., Pietruszka, M., Pałka, G.: Going full-TILT boogie on document understanding with text-image-layout transformer. In: Lladós, J., Lopresti, D., Uchida, S. (eds.) ICDAR 2021. LNCS, vol. 12822, pp. 732–747. Springer, Cham (2021). https://doi.org/10.1007/978-3-030-86331-9_47
29. Quirós, L., Vidal, E.: Reading order detection on handwritten documents. Neural Comput. Appl. **34**(12), 9593–9611 (2022). https://doi.org/10.1007/s00521-022-06948-5
30. Wang, J., Jin, L., Ding, K.: Lilt: A simple yet effective language-independent layout transformer for structured document understanding. In: Muresan, S., Nakov, P., Villavicencio, A. (eds.) Proceedings of the 60th Annual Meeting of the Association for Computational Linguistics (Volume 1: Long Papers), ACL 2022, Dublin, Ireland, 22–27 May 2022, pp. 7747–7757. Association for Computational Linguistics (2022). https://aclanthology.org/2022.acl-long.534
31. Wang, R., Fujii, Y., Popat, A.C.: Post-OCR paragraph recognition by graph convolutional networks. In: IEEE/CVF Winter Conference on Applications of Computer Vision, WACV 2022, Waikoloa, HI, USA, 3–8 January 2022, pp. 2533–2542. IEEE (2022). https://doi.org/10.1109/WACV51458.2022.00259
32. Wang, Z., Xu, Y., Cui, L., Shang, J., Wei, F.: Layoutreader: pre-training of text and layout for reading order detection. In: Moens, M., Huang, X., Specia, L., Yih, S.W. (eds.) Proceedings of the 2021 Conference on Empirical Methods in Natural Language Processing, EMNLP 2021, Virtual Event/Punta Cana, Dominican Republic, 7–11 November 2021, pp. 4735–4744. Association for Computational Linguistics (2021). https://doi.org/10.18653/v1/2021.emnlp-main.389
33. Xu, Y., et al.: Layoutlmv2: multi-modal pre-training for visually-rich document understanding. In: Zong, C., Xia, F., Li, W., Navigli, R. (eds.) Proceedings of the 59th Annual Meeting of the Association for Computational Linguistics and the 11th International Joint Conference on Natural Language Processing, ACL/IJCNLP 2021, (Volume 1: Long Papers), Virtual Event, 1–6 August 2021, pp. 2579–2591. Association for Computational Linguistics (2021). https://doi.org/10.18653/v1/2021.acl-long.201

34. Xu, Y., Li, M., Cui, L., Huang, S., Wei, F., Zhou, M.: Layoutlm: pre-training of text and layout for document image understanding. In: Gupta, R., Liu, Y., Tang, J., Prakash, B.A. (eds.) KDD 2020: The 26th ACM SIGKDD Conference on Knowledge Discovery and Data Mining, Virtual Event, CA, USA, 23–27 August 2020, pp. 1192–1200. ACM (2020). https://doi.org/10.1145/3394486.3403172
35. Zhong, X., Tang, J., Jimeno-Yepes, A.: Publaynet: largest dataset ever for document layout analysis. In: 2019 International Conference on Document Analysis and Recognition, ICDAR 2019, Sydney, Australia, 20–25 September 2019, pp. 1015–1022. IEEE (2019). https://doi.org/10.1109/ICDAR.2019.00166

TDAE: Text Detection with Affinity Areas and Evolution Strategies

Kefan Ma, Yuchen Luo, Zheng Huang$^{(\boxtimes)}$, Kai Chen, Jie Guo,
and Weidong Qiu

School of Electronic Information and Electrical Engineering,
Shanghai Jiao Tong University, Shanghai, China
{entropy2333,luoyuchen,huang-zheng,kchen,guojie,qiuwd}@sjtu.edu.cn

Abstract. Text detection in natural scenes has evolved considerably
in recent years. Segmentation-based methods are widely used for text
detection because they are robust to detect text of any shape. However, most previous works focus on word-level detection and neglect
the regions between adjacent words, which are helpful when some text
instances are very close. In this paper, we propose a novel image feature named affinity area that exploits the area between two adjacent
text instances to enhance the detection capability. We design an affinity
module to generate annotations based on existing word-level annotations
since no open dataset supports that. By optimizing this module, our
segmentation-based network TDAE can predict text regions and affinity
regions through which we can obtain the final detection results. Inspired
by the evolutionary strategy (ES), our network also utilizes an additional
novel fine-tuning step to update the parameters by adding adaptive but
random perturbations, which is quite different from the traditional gradient descent approach. Competitive results on ICDAR (2013, 2015, 2017),
CTW-1500, and SynthText benchmarks further demonstrate the effectiveness of TDAE.

Keywords: Affinity area · Text detection · Segmentation-based
network · Evolution strategy

1 Introduction

Text detection has always been a very popular topic for its wide applications. In
recent years, numerous methods [2,5,6,9,10,14,17,29] have shown competence
in text detection, which can be applied to face recognition, autonomous driving
and even artificial eyeballs for blind people.

These methods have switched from text instances initially in forms of quadrangles with fixed orientations like [11,28] to text instances with arbitrary shapes
or orientations like [18,25,29].

This work was supported by the National Natural Science Foundation of China (Grant
No. 92270201).

However, we notice that most models are only trained on word-level annotations and the output are all about text areas. As a result, models only focus on text regions and classify the rest of the image into the same class 'background', neglecting the space between two adjacent text instances.

CRAFT [1] introduces a new feature that labels the regions between two adjacent characters in the same word. It combines these areas with the predicted character regions to generate a complete word-level prediction. Since regions between characters have proven to be helpful in text detection, we believe that regions between adjacent words may also be useful and help the model achieve better generalization performance.

Unfortunately, no open dataset supports this new feature. To overcome this problem, we introduce an affinity module, which generates affinity annotations from existing word-level annotations. Unlike CRAFT [1], we label the area between two adjacent text instances instead of characters. After training with this additional, new feature, our network can effectively output predictions of text areas and affinity areas, which are combined to obtain the final results.

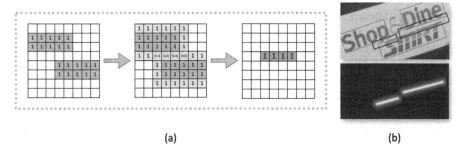

(a)   (b)

Fig. 1. In (a), the green areas represent text instances, while the yellow areas represent the part we just expanded. We mark the overlap of two yellow regions with a red boundary line, which is the affinity region we need. (b) shows a visualization of original word-level annotations, along with the labels of affinity regions. (Color figure online)

In addition, we add an innovative fine-tuning step to the training process. It can operate on parallel workers by generating noise to disrupt the model. And each noise added is adaptive to the parameters in the neural network layers. Since the perturbations are random and unpredictable, the model can decide whether to update the parameters based on the feedback from the evaluation.

The contributions of our paper can be described as follows:

1) As shown in Fig. 1, we introduce a new feature (affinity area) between adjacent text instances and propose a label generation method to serve it.
2) We propose TDAE, a segmentation-based network that is fully convolutional with two decoders (one for the text area and the other for the affinity area). TDAE is also equipped with a module to generate affinity annotations.

3) A novel, additional fine-tuning step is applied to help models previously trained with gradient descent methods to achieve better performance. To the best of our knowledge, we are the first to apply the evolution strategy (ES) to text detection.

4) Extensive experiments on five benchmark datasets demonstrate that our proposed method helps improve the detection capability. Compared with existing methods, TDAE can achieve competitive or state-of-the-art results with a better inference speed.

The rest of this paper is organized as follows. Section 2 briefly reviews related works. Section 3 gives details of the proposed method. Section 4 presents the experimental results, and Sect. 5 draws concluding remarks.

2   Related Works

Text detection has been well developed in recent years. We can roughly classify the existing methods into regression-based methods and segmentation-based methods.

Regression-Based Methods. There are already numerous object detectors or text detectors that use regression-based methods. Models using these methods first output multiple anchor points with different scales of width as candidates, and second pass them to the network, which scores each candidate. The score indicates the probability of the presence of any instance. This approach usually requires simple post-processing algorithms (e.g., non-maximal suppression also known as NMS). TextBoxes [11] modifies the convolution kernel and anchor boxes to efficiently capture text instances of various shapes, which addresses the problem that most text is presented in irregular bounding boxes of various aspect ratios. DMPNet [15] utilizes quadrilateral sliding windows to recall the text with higher overlapping area, and designed a sequential protocol for relative regression. Moreover, Mask R-CNN [4] shows satisfactory performance by combining Feature Pyramid Network, Region Proposal Network (RPN), and Fast R-CNN. In addition, EAST [29] will generate pixel-level text score maps for various channels and apply regression at the pixel level for multi-directional text detection. Unfortunately, most of them are limited when dealing with irregular shapes (e.g., curved shapes) because they cannot capture all possible shapes present in the image. Then TextSnake [17] appears, which can effectively render text instances with random shapes.

Segmentation-Based Methods. Segmentation-based text detection is another direction of text detection, which focuses directly on text regions at the pixel level. Text instances are detected by estimating the probability that a pixel is a tiny part of a text region. PSENet [24] applies multiple kernels to search for text regions on a reduced segmentation graph, and the segmentation region is continuously scaled up until the best result appears. Inspired by Mask R-CNN [4], TextSpotter [6] is an end-to-end segmentation-based model that includes both word-level and character-level segmentation. In CharNet [27], one branch locates the location of text regions, and the other locates and identifies characters. CRAFT [1] exploits the character-level annotations from SynthText [3] to generate character predictions and inter-character region predictions. It is firstly trained on SynthText and then fine-tuned on real-scene images through weakly supervised learning. SegLink [22] focuses on text grids, i.e., partial text fragments, and connects these fragments with additional link predictions. DB [12] is to increase the contribution of the bounding boxes of text instances to the training loss, allowing the model to better distinguish between the boundaries of text instances and core regions. In [18], candidate boxes are generated by sampling and grouping corner points, which are further scored by segmentation maps and suppressed by NMS.

3   Approach

In this section, we first introduce the affinity module and explain how the new annotations are generated. Next, we present the overall pipeline and details of TDAE. In the end, we illustrate the method of label generation and the algorithm for additional fine-tuning.

3.1   Affinity Module

In the detection process, we usually pay attention to the feature information of the text instances, such as shape, rotation, etc. However, the feature information of the area between text instances is often neglected, which is also a part of the entire image and has rich feature information. Therefore, we propose a module to make the most of image information, which can innovatively create annotations of affinity areas from existing word-level annotations.

In this module, we aim to equidistantly enlarge the polygons annotated in the image as shown in Fig. 1. Specifically, we first calculate the average height of word-level annotations. Then we expand the annotations by a certain percentage of the average height, and loop over all annotation pairs to generate new affinity annotations. The detailed processes are shown in Algorithm 1.

3.2   Label Generation

In contrast to the original annotations, we use segmentation masks with different shrinkage offsets to represent text instances and affinity areas in this paper. Inspired by the algorithm in [24], we first obtain an original polygon OP and calculate the shrinking offset s to get the shrunken polygon SP. Subsequently, SP is transformed into a 0/1 binary mask, which represents the segmentation labels in ground truth.

$$s = r \cdot A_{poly}(OP) \cdot (1 - f^2) \tag{1}$$

where $A_{poly}(\cdot)$ is a function to calculate the polygon area; r is the reciprocal of the polygon perimeter; f is a shrinkage factor which is set to 0.45.

Algorithm 1. Generation step in the affinity module

Require: An image in the dataset, I. Word-level annotations in I, A_i ($i = 1, 2, \cdots, n$).
   The ratio for polygon enlargement, R_{pe}.
Ensure: The list of affinity area annotations, $List\_a$.
 1: Calculate the average height of all A_i, H_a.
 2: Enlarge A_i by the offset $H_a * R_{pe}$.
 3: $List\_a = [\ ]$
 4: **for** i in $range(n)$ **do**
 5:     **for** j in $range(i + 1, n)$ **do**
 6:         $Canvas1 = zeros\_like(I.shape)$
 7:         $Canvas2 = zeros\_like(I.shape)$
 8:         $fillPoly(Canvas1, A_i, 1)$, elements in A_i are marked 1.
 9:         $fillPoly(Canvas2, A_j, 1)$, elements in A_j are marked 1.
10:         $Amap = Canvas1 + Canvas2$
11:         $Amap[C_i < 2] = 0$, C_i is the element of $Amap$.
12:         $Amap[C_i > 1] = 1$
13:         **if not** 1 in $Amap$ **then**
14:             **continue**
15:         **end if**
16:         $tmp\_area = minAreaRec(Amap)$
17:         $new\_pts = getContourPoints(tmp\_area)$
18:         $List\_a.append(new\_pts)$
19:     **end for**
20: **end for**
21: **return** $List\_a$

For affinity areas, we propose a novel label-making method. By using the previously mentioned method, we obtain the shrunken polygon SP from OP. As shown in Fig. 2b, SP is further clipped to get the ridge R. Figure 1b shows an visualization of this feature. The element E in an affinity label can be calculated as follows:

$$E_{(i,j)} = \begin{cases} 1 & (i,j) \in R \\ 0 & (i,j) \in OP - SP \\ 1 - \frac{\min(D_m s)}{L} & (i,j) \in SP - R \end{cases} \tag{2}$$

where (i, j) represents the coordinate of an element; $SP - R$ can be defined as the area in SP but not in R; D_m ($m = 1, 2, \cdots, n$) are the vertical distances from a pixel to each side of the ridge area and L is the maximum shrinkage offset from SP to R.

3.3   Network Design

In this paper, we employ ResNet as the backbone, and adopt a FPN [13] style structure for multi-scale feature fusion. TDAE will produce a feature map consisting of four different levels of upsampling concatenated. Figure 3 illustrates the overall pipeline of TDAE.

$$F_{fused} = \text{Concat}\left(F_1, U_2(F_2), U_4(F_3), U_8(F_4)\right) \tag{3}$$

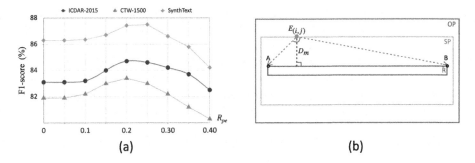

Fig. 2. (a) Ablation study on area expansion offset. (b) The label-making process of affinity areas.

where F_{fused} is the fused features; F_i ($i = 1, 2, 3, 4$) are feature maps from ResNet in different scales; $U_j(\cdot)$ indicates up-sampling function and the subscript of $U_j(\cdot)$ is the scale of up-sampling.

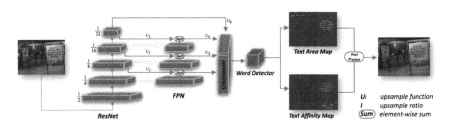

Fig. 3. The structure of our proposed network. The word detector utilizes stacked deformable convolution layers to generate maps of text areas and the affinity areas. The fractions in the figure are the ratio of original input image.

In the end, F_{fused} will be fed into a word-level detector. The detector consists of a $Conv(3 \times 3)$-BN-$ReLU$ layer, a $ConvTranspose$-BN-$ReLU$ layer, a $Convtranspose$ layer and finally a $Sigmoid$ layer. Upon receiving F_{fused}, the detector will generate an affinity area map M_a and a text core area map M_c, as depicted in Fig. 3. Those maps will be utilized to obtain the final score map M_s, which can be summeried as follows: according to the pre-set threshold, which is 0.21 in this paper, M_a is initially transformed into a mask, labeling affinity areas as 0 and the rest as 1. Next, we apply it to the M_c to filter some none-text regions with relatively high confidence. At last, the final result is generated.

3.4 Additional Fine-Tuning Step

We explore a new approach to optimize the model, which is different from the traditional gradient descent algorithm. Inspired by [21], we propose an additional fine-tuning step, which is a method based on the classic black-box optimization algorithm, to help a previously trained model obtain better performance. In particular, there exists no need for backpropagation.

Specifically, multiple candidate models are obtained in an iteration by perturbing the internal parameters of the original model with different noises. Then we calculate the reward for every single perturbed model. Finally, the model parameters are updated to better internal parameters based on the rewards obtained from the evaluation.

Algorithm 2. Additional fine-tuning step for TDAE

Require: The TDAE model, M. The noise intensity, σ. Parameters to be updated in a model, θ. The training rounds, T. The perturbation numbers per round (also known as the number of workers), N. The evaluation function, Eval.

Ensure: The updated model, M.

1: Set $Reward = 0$
2: **for** i in T **do**
3:     **for** j in N **do**
4:         Make a copy of M, M_c
5:         **for** θ in M_c **do**
6:             $\epsilon \sim N(0, I)$ with $seed_j$
7:             $\epsilon_a = \epsilon - \text{mean}(\epsilon) + \sigma \cdot \text{mean}(\theta)$
8:             $\theta = \theta + \epsilon_a$
9:         **end for**
10:         Obtain the perturbed model, M_p
11:         $reward_j = \text{Eval}(M_p)$
12:         Collect all $seed_j$ and $reward_j$
13:     **end for**
14:     Get the seed $Best$ with the largest reward, R_{max}
15:     **if** $R_{max} < Reward$ **then**
16:         **Continue**
17:     **end if**
18:     $Reward = R_{max}$
19:     **for** θ in M **do**
20:         $\epsilon \sim N(0, I)$ with $Best$
21:         $\epsilon_a = \epsilon - \text{mean}(\epsilon) + \sigma \cdot mean(\theta)$
22:         $\theta = \theta + \epsilon_a$
23:     **end for**
24: **end for**
25: **return** M

However, the generation of perturbations must be carefully designed due to the different shapes and average values of the parameters in each layer. If the perturbation is not appropriate, the robustness of the model tends to fail.

To solve this problem well, we propose an adaptive perturbation generation method. The main idea can be summarized in Algorithm 2. Moreover, a task under this algorithm can be applied to parallel workers. All workers know the perturbation seeds, and only need to exchange one scalar to reach an agreement on parameter updates.

3.5 Loss Function

In order to detect text core area and text affinity area simultaneously, it requires taking text area loss L_c and affinity area loss $\mathcal{L}_a$ both into consideration during training. The overall training loss can be formulated as:

$$\mathcal{L} = \mathcal{L}_a + \lambda_1 \mathcal{L}_c \tag{4}$$

$$\mathcal{L}_c = F_{l1}(X_1, Y_1) + \lambda_2 * F_{bce}(X_1, Y_1) + F_{dice}(X_1, Y_1) \tag{5}$$

$$\mathcal{L}_a = F_{l1}(X_2, Y_2) + \lambda_3 * F_{bce}(X_2, Y_2) \tag{6}$$

where λ_i $(i = 1, 2, 3)$ are weighted values to balance these tasks. F means loss function with subscript indicating the specific loss type; X_1 and X_2 are outputs from the detector; Y_1 and Y_2 are maps generated from text affinity and text area annotations respectively. To be more specific, we apply L1 distance and binary cross entropy (BCE) loss for both $\mathcal{L}_c$ and $\mathcal{L}_a$. And Dice loss [23] is adopted for $\mathcal{L}_c$ to mitigate data imbalance problems.

4 Experiments

In this section, we first conduct ablation studies for TDAE on three challenging public benchmarks: ICDAR 2015, CTW-1500 and SynthText. Then we compare TDAE with previous state-of-the-art methods on the ICDAR (2013, 2015, 2017) and CTW-1500 dataset.

4.1 Datasets

There are five datasets utilized in this work. The details are as follows:

ICDAR-2013 [8]. It is a focused scene text dataset in English, introduced during the ICDAR 2013 Robust Reading Competition. The training part consists of 229 images and the testing part consist of 233 images. The annotations are labeled with rectangular boxes in word-level.

ICDAR-2015 [7]. This dataset is created by ICDAR 2015 Robust Reading Competition. It is mainly composed of two parts: 1000 images for the training part and 500 for the testing part.

ICDAR-2017 [19]. It contains 7,200 training images, 1,800 validation images, and 9,000 testing images with texts in 9 languages for multi-lingual scene text detection.

SynthText [3]. Its available resources consist of 800,000 images and there are approximately 8 million instances of synthetic words. Each instance is annotated with word-level and char-level bounding boxes.

CTW-1500 [16]. It is composed of 1000 training images and 500 for testing. Each image has curved text instances, which are annotated by polygons with 14 vertices. The dataset contains a lot of horizontal and multi-oriented text.

4.2 Implementation Details

We implement TDAE with PyTorch and conduct experiments on two NVIDIA Tesla M40 12GB. Following the previous work [12], our training procedure consists of two steps: we first pre-train the network on SynthText [3] for 50k iterations, then each benchmark dataset is adopted for fine-tuning, after which we implement the additional fine-tuning step. The training batch size is set to 8. Stochastic gradient descent (SGD) optimizer is adopted in the training process. The weight decay is set to 0.0001, momentum is set to 0.9. The initial learning rate for pre-training and fine-tuning is set to $7e^{-3}$ and $3.5e^{-3}$. The number of epochs for fine-tuning is set to 700. In the training, we define $\lambda_1 = 7$ and $\lambda_2 = \lambda_3 = 0.1$ respectively. Further, basic data augmentation methods like crops, flips, and rotations are applied. All the images are resized to 640×640 in the experiments for better training efficiency.

For additional fine-tuning step, the noise intensity σ is set to 0.007; the standard deviation I is set to 0.01 in the normal distribution. Moreover, the number of children N, also called perturbation number, is 6 and the total round is set to 800.

4.3 Ablation Study

We conduct ablation studies on **ICDAR-2015, CTW-1500** and **SynthText** to analyze the effectiveness of TDAE with the new feature, fine-tuning step and different backbones. The detailed experimental results are shown in Table 1.

Table 1. Comparison to the traditional semantic segmentation baseline with the same backbone. AA and AFS refer to Affinity Area and Additional Fine-tune Step respectively. **P**, **R**, and **F** represent Precision, Recall, and F-measure, respectively.

Model	ICDAR-2015			CTW-1500			SynthText		
	P	R	F	P	R	F	P	R	F
TDAE (ResNet-50)	90.0	81.9	85.8	86.4	80.3	83.2	93.7	83.3	88.2
w/o AA	87.5	80.2	83.7	84.0	78.0	80.9	92.5	82.8	87.4
w/o AFS	88.3	81.4	84.7	86.0	79.3	82.6	93.4	82.2	87.5
w/o AA & AFS	86.7	80.0	83.1	82.9	77.5	80.1	92.3	81.1	86.3
TDAE (ResNet-18)	85.3	79.7	82.4	84.3	77.4	80.7	91.7	81.7	86.4
w/o AA	83.7	77.9	80.7	82.2	74.9	78.4	90.5	78.9	84.3
w/o AFS	85.2	79.2	82.1	83.9	76.8	80.2	91.4	80.5	85.6
w/o AA & AFS	83.4	77.2	80.2	82.1	74.1	77.9	89.8	78.7	83.9

Affinity Area (AA). From Table 1, we can see that our proposed new feature improves the performance significantly for both ResNet-18 and ResNet-50 on the three datasets, compared with models that do not take AA into training.

For the ResNet-18 backbone, affinity area achieves 1.9% (ICDAR-2015), 2.3% (CTW-1500), 1.7% (SynthText) performance gain in terms of the F-measure. For the ResNet-50 backbone, affinity area brings 1.6% (ICDAR-2015), 2.5% (CTW-1500), 1.2% (SynthText) improvements, respectively. Figure 4 illustrates the difference of outputs from the model trained with AA and the one without it. These results validate that using affinity area as an extra feature can effectively help models to segment accurately, especially in areas with dense text instances.

Fig. 4. Some visualization results of ablation study on the affity area. (a) is the prediction from the traditional semantic segmentation baseline. (b) is the prediction of TDAE. (c) and (d) are the probability maps of text area and affinity area from TDAE.

Additional Fine-Tuning Step (AFS). As shown in Table 1, AFS can also brings 0.4-1.1% performance gain since suitable perturbation can improve the performance of models. For ICDAR-2015, 0.5% (with ResNet-18) and 0.6% (with ResNet-50) improvements are achieved by AFS. For CTW-1500, 0.5% (with ResNet-18) and 0.8% (with ResNet-50) increments are obtained by AFS. And models on SynthText gain 0.4% (with ResNet-18) and 1.1% (with ResNet-50) through AFS. Thus, our proposed fine-tuning policy also contributes to the satisfactory performance of the model.

How We Define Two Instances Are Adjacent. To answer the question, we investigate the effect of the offset for polygon expansion in TDAE. The models are evaluated on the ICDAR-2015, CTW-1500 and SynthText datasets. According to Algorithm 1, we obtain the average height H_a from all instances in an image. Then we multiply H_a with a ratio R_{pe} as the offset of polygon expansion. In details, we design a series of experiments to observe the performance of the model under different R_{pe} from 0 to 0.4. And we can find in Fig. 2a that when R_{pe} is too large, affinity areas become large as well, which will weaken the detection of the text area in the post-process. However when R_{pe} is too small, the affinity area is too tiny to learn in most images. Therefore, feature information from the text affinity map is insufficient, which can hardly help the model predict. In the end, we set the proper R_{pe} to 0.2 and apply it to ICDAR-2013, ICDAR-2017. Statistics from Table 1 prove that this value still works on other datasets for the new feature.

4.4   Comparisons with Previous Methods

We compare TDAE with previous state-of-the-art methods. And the evaluation is performed on four standard benchmarks, including one benchmark for near-horizontal texts, one benchmark for multi-oriented texts, one benchmark for curved text detection, and one multi-language benchmark for long text lines.

Table 2. Model performance on ICDAR-2013 and CTW-1500 datasets. FPS is only for reference since the experimental environments are different with ICDAR-2013 unincluded. We report the best FPSs, each of which was reported in the original paper. **P**, **R**, and **F** represent Precision, Recall, and F-measure, respectively.

Model	ICDAR-2013			CTW-1500			FPS
	P	R	F	P	R	F	
EAST [29]	-	-	-	78.7	49.1	60.4	13.2
PixelLink [2]	88.6	87.5	88.1	-	-	-	-
TextSpotter [6]	88	87	88	-	-	-	-
SSTD [5]	89	86	88	-	-	-	7.7
PSENet [24]	-	-	-	84.8	79.7	82.2	12.4
Lyu et al. [18]	92.0	84.4	88.0	-	-	-	5.7
DB (ResNet-50) [12]	-	-	-	86.9	80.2	83.4	22
CoutourNet [26]	-	-	-	84.1	**83.7**	83.9	4.5
TDAE (ResNet-50)	91.1	88.6	89.8	86.4	80.3	83.2	**26.2**
TDAE (ResNet-101)	**93.2**	**91.6**	**92.4**	**88.3**	81.5	**85.4**	12.5

Near-Horizontal Text Detection. First, we evaluate the effectiveness of TDAE in detecting near-horizontal text on the ICDAR 2013 dataset. As presented in the first column of Table 2, TDAE with the backbone of ResNet-50 and ResNet-101 both achieve very outstanding results. And the F-measure are 89.8% and 92.4% respectively, which outperform all previous works.

Curved Text Detection. To prove our method's effectiveness on curved text, we conduct experiments on the popular CTW-1500 dataset. The results of TDAE and comparison with previous methods are presented in the second column of Table 2. Our ResNet-50 and ResNet-101 both achieve a competitive or even state-of-the-art performance with a fast inference speed. TDAE (ResNet-101) outperforms the current best method by 1.5% in terms of F-measure.

Multi-oriented Text Detection. Moreover, we evaluate our method for multi-oriented text on the ICDAR-2015 dataset, which consists of many small, low-resolution, and long text instances. As shown in Table 3, our model (ResNet-50) outperforms previous state-of-the-art method by 0.4%. Compared with the former CNN-based methods, our proposed method achieves the best performance 85.8% in terms of F-measure.

Table 3. Detection results on ICDAR-2015 and ICDAR-2017 datasets. **P**, **R**, and **F** represent Precision, Recall, and F-measure, respectively.

Model	ICDAR-2015			ICDAR-2017			FPS
	P	R	F	P	R	F	
EAST [29]	83.6	73.5	78.2	-	-	-	13.2
PixelLink [2]	82.9	81.7	82.3	-	-	-	-
TextSpotter [6]	84	83	83	-	-	-	-
SSTD [5]	73	80	77	-	-	-	7.7
FOTS [14]	88.8	82.0	85.3	79.5	57.5	66.7	23.9
PSENet [24]	86.9	**84.5**	85.7	73.8	68.2	70.9	12.4
Lyu *et al.* [18]	89.5	79.7	84.3	74.3	70.6	72.4	3.6
DB (ResNet-50) [12]	88.2	82.7	85.4	**83.1**	67.9	74.7	26
Raisi *et al.* [20]	89.8	78.3	83.7	84.8	63.2	72.4	-
TDAE (ResNet-50)	90.0	81.9	85.8	77.3	70.8	73.9	**28.3**
TDAE (ResNet-101)	**92.4**	83.1	**87.5**	80.3	**71.9**	**75.9**	13.7

Fig. 5. Example visualization results on scene text detection.

Multi-lingual Text Detection. Finally, to demonstrate the robustness of our model for different languages, experiments are conducted on the large-scale scene text dataset ICDAR-2017. Results in Table 3 shown that, our proposed TDAE (ResNet-50) obtains a comparable detection performance with a better inference speed (28.3 vs. 26). When the backbone is ResNet-101, a more compelling result can be achieved (F-measure: 75.9%), outperforming all the other competitors by at least 1.2%.

Therefore, from these experimental results on ICDAR 2013, ICDAR 2015, Total-Text and CTW-1500, our proposed TKDE achieves competitive or state-of-the-art performance. In addition, TDAE can also inference at a comparatively fast speed (28.3 FPS on ICDAR-2015 dataset), which has a degree of ascendancy compared to some previous methods, due to the efficient post-processing. Some visualized examples using TDAE are represented in Fig. 5.

5   Conclusion and Future Work

In this paper, we propose a novel network that generates annotations for affinity areas in images, incorporates them into training, and applies an adaptive evolution strategy for model fine-tuning. Extensive experiments demonstrate that our proposed method can achieve comparable or state-of-the-art performance with a better inference speed. It is worth pointing out that our approach is flexible and adaptive, and can be applied to the recently popular Vision Transformer.

There are still multiple directions to explore in the future. First, we would like to create a dataset containing dense text instances, such as pages in a book, and optimize our network under this specific condition. Second, the additional fine-tuning algorithm can be introduced to the general instance-level segmentation tasks, setting more children for perturbations with more GPU resources.

References

1. Baek, Y., Lee, B., Han, D., Yun, S., Lee, H.: Character region awareness for text detection. In: CVPR, pp. 9365–9374. Computer Vision Foundation/IEEE (2019)
2. Deng, D., Liu, H., Li, X., Cai, D.: Pixellink: detecting scene text via instance segmentation. In: AAAI, pp. 6773–6780. AAAI Press (2018)
3. Gupta, A., Vedaldi, A., Zisserman, A.: Synthetic data for text localisation in natural images. In: CVPR, pp. 2315–2324. IEEE Computer Society (2016)
4. He, K., Gkioxari, G., Dollár, P., Girshick, R.B.: Mask R-CNN. In: ICCV, pp. 2980–2988. IEEE Computer Society (2017)
5. He, P., Huang, W., He, T., Zhu, Q., Qiao, Y., Li, X.: Single shot text detector with regional attention. In: ICCV, pp. 3066–3074. IEEE Computer Society (2017)
6. He, T., Tian, Z., Huang, W., Shen, C., Qiao, Y., Sun, C.: An end-to-end textspotter with explicit alignment and attention. In: CVPR, pp. 5020–5029. Computer Vision Foundation/IEEE Computer Society (2018)
7. Karatzas, D., et al.: ICDAR 2015 competition on robust reading. In: ICDAR, pp. 1156–1160. IEEE Computer Society (2015)
8. Karatzas, D., et al.: ICDAR 2013 robust reading competition. In: ICDAR, pp. 1484–1493. IEEE Computer Society (2013)
9. Liao, M., Lyu, P., He, M., Yao, C., Wu, W., Bai, X.: Mask textspotter: an end-to-end trainable neural network for spotting text with arbitrary shapes. IEEE Trans. Pattern Anal. Mach. Intell. **43**(2), 532–548 (2021)
10. Liao, M., Shi, B., Bai, X.: Textboxes++: a single-shot oriented scene text detector. IEEE Trans. Image Process. **27**(8), 3676–3690 (2018)
11. Liao, M., Shi, B., Bai, X., Wang, X., Liu, W.: Textboxes: a fast text detector with a single deep neural network. In: AAAI, pp. 4161–4167. AAAI Press (2017)
12. Liao, M., Wan, Z., Yao, C., Chen, K., Bai, X.: Real-time scene text detection with differentiable binarization. In: AAAI, pp. 11474–11481. AAAI Press (2020)
13. Lin, T., Dollár, P., Girshick, R.B., He, K., Hariharan, B., Belongie, S.J.: Feature pyramid networks for object detection. In: CVPR, pp. 936–944. IEEE Computer Society (2017)

14. Liu, X., Liang, D., Yan, S., Chen, D., Qiao, Y., Yan, J.: FOTS: fast oriented text spotting with a unified network. In: CVPR, pp. 5676–5685. Computer Vision Foundation/IEEE Computer Society (2018)
15. Liu, Y., Jin, L.: Deep matching prior network: toward tighter multi-oriented text detection. In: CVPR, pp. 3454–3461. IEEE Computer Society (2017)
16. Liu, Y., Jin, L., Zhang, S., Zhang, S.: Detecting curve text in the wild: new dataset and new solution. CoRR abs/1712.02170 (2017)
17. Long, S., et al.: TextSnake: a flexible representation for detecting text of arbitrary shapes. In: Ferrari, V., Hebert, M., Sminchisescu, C., Weiss, Y. (eds.) ECCV 2018. LNCS, vol. 11206, pp. 19–35. Springer, Cham (2018). https://doi.org/10.1007/978-3-030-01216-8_2
18. Lyu, P., Yao, C., Wu, W., Yan, S., Bai, X.: Multi-oriented scene text detection via corner localization and region segmentation. In: CVPR, pp. 7553–7563. Computer Vision Foundation/IEEE Computer Society (2018)
19. Nayef, N., et al.: ICDAR2017 robust reading challenge on multi-lingual scene text detection and script identification - RRC-MLT. In: ICDAR, pp. 1454–1459. IEEE (2017)
20. Raisi, Z., Naiel, M.A., Younes, G., Wardell, S., Zelek, J.S.: Transformer-based text detection in the wild. In: CVPR Workshops, pp. 3162–3171. Computer Vision Foundation/IEEE (2021)
21. Salimans, T., Ho, J., Chen, X., Sutskever, I.: Evolution strategies as a scalable alternative to reinforcement learning. CoRR abs/1703.03864 (2017)
22. Shi, B., Bai, X., Belongie, S.J.: Detecting oriented text in natural images by linking segments. In: CVPR, pp. 3482–3490. IEEE Computer Society (2017)
23. Sudre, C.H., Li, W., Vercauteren, T., Ourselin, S., Jorge Cardoso, M.: Generalised dice overlap as a deep learning loss function for highly unbalanced segmentations. In: Cardoso, M.J., et al. (eds.) DLMIA/ML-CDS -2017. LNCS, vol. 10553, pp. 240–248. Springer, Cham (2017). https://doi.org/10.1007/978-3-319-67558-9_28
24. Wang, W., et al.: Shape robust text detection with progressive scale expansion network. In: CVPR, pp. 9336–9345. Computer Vision Foundation/IEEE (2019)
25. Wang, W., et al.: Efficient and accurate arbitrary-shaped text detection with pixel aggregation network. In: ICCV, pp. 8439–8448. IEEE (2019)
26. Wang, Y., Xie, H., Zha, Z., Xing, M., Fu, Z., Zhang, Y.: Contournet: taking a further step toward accurate arbitrary-shaped scene text detection. In: CVPR, pp. 11750–11759. Computer Vision Foundation/IEEE (2020)
27. Xing, L., Tian, Z., Huang, W., Scott, M.R.: Convolutional character networks. In: ICCV, pp. 9125–9135. IEEE (2019)
28. Zhang, Z., Shen, W., Yao, C., Bai, X.: Symmetry-based text line detection in natural scenes. In: CVPR, pp. 2558–2567. IEEE Computer Society (2015)
29. Zhou, X., et al.: EAST: an efficient and accurate scene text detector. In: CVPR, pp. 2642–2651. IEEE Computer Society (2017)

Visual Information Extraction in the Wild: Practical Dataset and End-to-End Solution

Jianfeng Kuang[1], Wei Hua[1], Dingkang Liang[1], Mingkun Yang[1], Deqiang Jiang[2], Bo Ren[2], and Xiang Bai[1(✉)]

[1] Huazhong University of Science and Technology, Wuhan, China
{kuangjf,whua_hust,dkliang,yangmingkun,xbai}@hust.edu.cn
[2] Tencent YouTu Lab, Shanghai, China
{dqiangjiang,timren}@tencent.com

Abstract. Visual information extraction (VIE), which aims to simultaneously perform OCR and information extraction in a unified framework, has drawn increasing attention due to its essential role in various applications like understanding receipts, goods, and traffic signs. However, as existing benchmark datasets for VIE mainly consist of document images without the adequate diversity of layout structures, background disturbs, and entity categories, they cannot fully reveal the challenges of real-world applications. In this paper, we propose a large-scale dataset consisting of camera images for VIE, which contains not only the larger variance of layout, backgrounds, and fonts but also much more types of entities. Besides, we propose a novel framework for end-to-end VIE that combines the stages of OCR and information extraction in an end-to-end learning fashion. Different from the previous end-to-end approaches that directly adopt OCR features as the input of an information extraction module, we propose to use contrastive learning to narrow the semantic gap caused by the difference between the tasks of OCR and information extraction. We evaluate the existing end-to-end methods for VIE on the proposed dataset and observe that the performance of these methods has a distinguishable drop from SROIE (a widely used English dataset) to our proposed dataset due to the larger variance of layout and entities. These results demonstrate our dataset is more practical for promoting advanced VIE algorithms. In addition, experiments demonstrate that the proposed VIE method consistently achieves the obvious performance gains on the proposed and SROIE datasets. The code and dataset will be available at https://github.com/jfkuang/CFAM.

Keywords: Visual Information Extraction · Document Understanding · Document Semantics Extraction

---

J. Kuang and W. Hua—Equal contribution.
Work done when Wei Hua was an intern at Tencent.

© The Author(s), under exclusive license to Springer Nature Switzerland AG 2023
G. A. Fink et al. (Eds.): ICDAR 2023, LNCS 14192, pp. 36–53, 2023.
https://doi.org/10.1007/978-3-031-41731-3_3

1   Introduction

Visual Information Extraction (VIE), automatically extracting structured information from visually-rich document images, is an essential step towards document intelligence [9,31,42]. It involves extracting values of entities from images and reasoning about their relations, which is a challenging cross-domain problem that requires both visual and textual understanding, such as table comprehension and document analysis [10,29,32,39]. This task becomes more difficult in real-world scenarios where diverse layouts, noisy backgrounds, and large variances of entities exhibit in the wild.

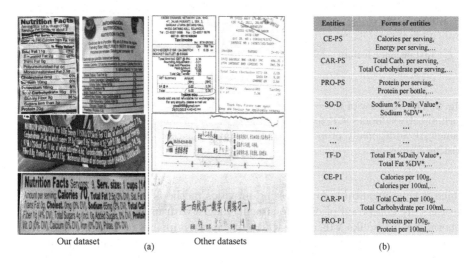

Fig. 1. (a) exhibits a few typical examples from our dataset and others. (b) shows some typical entities and their different forms in our dataset. Note that the actual entity name is very long, so we use the abbreviation of the entity name. More details of our POIE dataset can be found at https://github.com/jfkuang/CFAM

Table 1. Statistics of the representative VIE datasets and our proposed dataset.

Dataset	Year	Number of images			Number of entities	Number of instance	Language	Scene Type
		Train	Test	Total				
SROIE [7]	ICDAR2019	626	347	973	4	52,316	EN	Receipt
EPHOIE [30]	AAAI2021	1,183	311	1,494	10	15,771	CN	Examination Paper
POIE (Ours)	-	2,250	750	3,000	21	111,155	EN	Product

Datasets play an essential role in data-driven problems. In the document VIE area, a number of datasets [2,7,21,30], have been proposed. Specifically, SROIE [7] is the most widely used dataset, in which the images are scanned receipts printed in English. EPHOIE [30] is collected for Chinese document VIE,

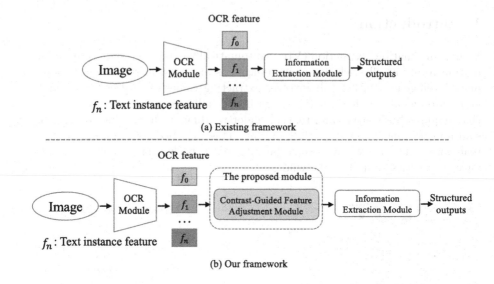

Fig. 2. The overview of the existing framework and our framework. The existing framework directly adopts OCR features as the input of an information module, while our framework additionally proposes a contrast-guided feature adjustment module to the existing framework.

which is thus unsuitable for assessing previous methods for English VIE. These datasets are composed of only document images with moderate scale. For more general scenarios, like VIE in the wild, the captured images have larger variances in layouts, backgrounds, and fonts, as well as various entities. Consequently, the available datasets cannot fully reveal the challenges of real-world applications.

To better study VIE in the wild, we collect a large-scale challenging dataset called **P**roducts for **OCR** and **I**nformation **E**xtraction (POIE), which consists of camera images from Nutrition Facts label of products in English. Compared with existing VIE datasets [7,30], POIE has several distinct characteristics. First, it is the largest VIE dataset with high-quality annotations, where 3,000 images with 111,155 text instances are collected. As indicated in Table 1, it is over 2× larger than the previous largest public VIE dataset. Second, POIE is particularly challenging, where the images are in diverse layouts and distorted with various folds, bends, deformations as well as perspectives. We show some typical examples in Fig. 1(a). Third, POIE has up to 21 kinds of entities, some of which come in numerous forms. For example, in Fig. 1(b), several forms of some entities are given. It is quite common in the real-world applications that each entity presents in various forms, demonstrating VIE in the wild is more difficult.

We observe that previous end-to-end approaches [30,40] achieve lower performance on POIE, especially on the information extraction task, due to complex layouts and variable entities. Besides, these methods directly adopt OCR features as the input of the following information extraction module. We argue that the reason is there is a severe semantic gap between the tasks of OCR and information extraction while directly feeding OCR features into the following information extraction module. Recently, we have witnessed the rise of contrastive learning in computer vision, which has received extensive attention in various visual tasks [1,6,12,24,37] while has rarely been noticed in VIE tasks. In this paper, we propose a novel end-to-end framework for VIE. We adopt contrastive learning to effectively establish the connections between the tasks of OCR and information extraction. Specifically, the key component of our framework is a plug-and-play Contrast-guided Feature Adjustment Module (CFAM), as shown in Fig. 2. In CFAM, we design the feature representation for OCR and information extraction (i.e., instance features and entity features for OCR and information extraction). As a result, CFAM constructs a similarity matrix reflecting the relations between entity features and instance features to adjust the instance features more appropriately for the following information extraction task.

In summary, the main contributions of this paper are two-fold: 1) we propose a large-scale dataset consisting of camera images with variable layouts, backgrounds, fonts, and much more types of entities for VIE in the wild. 2) we design a novel end-to-end framework with a plug-and-play CFAM for VIE, which adopts contrastive learning to narrow the semantic gap caused by the difference between the tasks of OCR and information extraction.

2 Related Works

2.1 VIE Datasets

Currently, a few datasets [2,7,21,26,30] are proposed for VIE on document images. Specifically, SROIE [7] is the most widely used dataset, which significantly promotes the development of this field. The images of SROIE are scanned receipts in printed English. Each image is associated with complete OCR annotations and the values of four key text fields. Besides, EPHOIE [30] is collected for Chinese document VIE, where the images have complex backgrounds and diverse text styles. The details of these datasets mentioned above are shown in Table 1. We can observe that these existing datasets are composed of only document images with moderate scale. To better explore VIE in the wild, a large-scale dataset with more entities and larger variances in layouts, backgrounds, and fronts is urgently required.

2.2 VIE Methods

According to the pipeline of VIE, existing approaches [2,16,27,34,38] can be divided into two categories: methods in two-stage and end-to-end paradigms. Most works with a two-stage pipeline focus on the second stage for information extraction. In these methods, the OCR results are first obtained via an individual OCR extractor. In Post-OCR parsing [8], the coordinates of text bounding boxes are applied during the second stage. To better model the layout structure and visual cues of documents, LayoutLM [34] employed a pre-training strategy inspired by BERT. GraphIE [23] and PICK [38] constructed a graph according to the OCR results and applied Graph Neural Networks (GNNs) to extract the global representation for further improvement. In CharGrid [11], CNN is used to integrate semantic clues and the layout information. MatchVIE [27] noticed the importance of modeling the relationships between entities and text instances. However, it required additional annotations of all key-value pairs. All of these methods concentrated on the context modeling between OCR results in the second stage but ignored the accumulative errors from the preceding OCR module.

Recently, an increasing number of VIE methods have been proposed in an end-to-end fashion. EATEN [3] first generated feature maps from input images and attached several entity-aware decoders to predict all entities. However, it can only handle documents with a fixed layout. TRIE [40] was an end-to-end trainable framework to solve the VIE task, which focused more on the performance of entity extraction. VIES [30] improved each part of VIE, like text detection, recognition, and information extraction, but incurred obvious costs. Donut [13] can directly extract information from the input images without text spotting, while it needed massive data for pre-training. All of the above methods directly took OCR features as the input of the following information extraction module. Different from the previous end-to-end approaches, we propose a novel framework in an end-to-end manner to establish the connections between the tasks of OCR and information extraction.

2.3   Contrastive Learning

Contrastive learning [12], a typical way for visual representation learning, has attracted lots of attention in many fields and obtained great progress [1, 24, 36]. Contrastive learning allows samples of positive pairs to lie close together in the latent space, while samples belonging to negative pairs are repelled in the latent space. CLIP [24] is a representative work for vision-language pre-training via contrastive learning on image-text pairs. Since then, plenty of multi-modal contrastive learning methods have been proposed [15, 22, 25, 28, 35, 41]. However, to the best of our knowledge, contrastive learning has not been thoroughly studied in the VIE field. In this paper, we propose to use contrastive learning to supervise the construction of relations between the tasks of OCR and information extraction. Different from the above approaches, where the image features and text features are taken as inputs, only the features of text instances are fed into the information extraction module in the VIE task. Therefore, how to represent OCR features as well as entity features, and construct the relations between them to narrow the semantic gap should be of great importance for end-to-end VIE.

3   POIE Dataset

3.1   Data Collection

Products for **O**CR and **I**nformation **E**xtraction (POIE) dataset derives from camera images of various products in the real world. The images are carefully selected and manually annotated. Our labeling team consists of 8 experienced labelers. We first crop the nutrition tables from product images and adopt multiple commercial OCR engines (Azure and Baidu OCR) for pre-labeling. Then we use LabelMe[1] to manually check the annotation of the location as well as transcription of every text box, and the values of entities for all the text in the images and repaired the OCR errors found. After discarding low-quality and blurred images, we obtain 3,000 images with 111,155 text instances.

3.2   Data Characteristics

To the best of our knowledge, POIE is the largest dataset with both OCR and VIE annotations for the end-to-end VIE in the wild. The images in POIE contain Nutrition Facts labels from various commodities in the real world, which have larger variances in layout, severe distortion, noisy backgrounds, and more types of entities than existing datasets. The comparison of dataset statistics is shown in Table 1. Existing datasets mainly consist of document images with insufficient diversities of layout, background disturbs, and entity categories. Therefore, they cannot fully illustrate the challenges of some practical applications, like VIE on the Nutrition Facts label. Compared with these datasets, POIE contains images with variable appearances and styles (such as structured, semi-structured, and unstructured styles), complex layouts, and noisy backgrounds distorted by folds, bends, deformations, and perspectives (typical examples are shown in Fig. 1(a)). Moreover, the types of entities in POIE reach 21, and a few entities have different forms (some typical entities with various forms are shown in Fig. 1(b)), which is very common and pretty challenging for VIE in the wild. Besides there are often multiple words in each entity, which appears zero or once in every image. These properties mentioned above can help enhance the robustness and generalization of VIE models to better cope with more challenging applications.

3.3   Data Split and Evaluation Protocol

POIE is divided into training and testing sets, with 2,250 and 750 images, respectively. We use the performance of detection (DET), recognition (REC), and information extraction (IE) in the end-to-end pipeline to evaluate all methods on the proposed POIE. Following the settings of SROIE [7] and EPHOIE [30], F1-Score is applied as the evaluation metric for the three tasks.

---

[1] https://github.com/wkentaro/labelme.

4   Our Method

The overall pipeline of our method is illustrated in Fig. 3. Different from other methods, we introduce contrastive learning into the end-to-end trainable framework for effectively bridging the modules of OCR and information extraction. Specifically, it is composed of a text detector, a Contrast-guided Feature Adjustment Module (CFAM), a recognizer module, and an information extraction module. Given an input image,

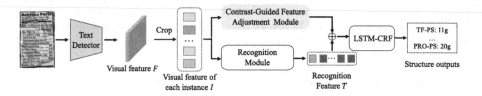

Fig. 3. The overall framework of our method. It consists of a text detector, a contrast-guided feature adjustment module, a recognition module, and an information extraction module.

- The CNN-based text detector not only localizes all text instances but also yields visual feature maps F for subsequent modules. The text detector detects at the line level.
- With RoIAlign, the visual features of each instance I are obtained from F. Then, I is simultaneously fed into CFAM and a recognition module to generate a similarity matrix S and the recognition features T. In the recognition module which is two-layers transformer decoder, we use the intermediate features in transformer as recognition features T and obtain the recognition results of all text. In the CFAM, I is first encoded to E. Meanwhile, we use a serial of learnable vectors as the entity features L'. Subsequently, the pair-wise similarity between L' and E is calculated.
- Finally, the recognition features T, the encoded features E, and similarity matrix S are added, and the results are fed into the following information extraction module (LSTM-CRF) to predict the structural outputs.

4.1   Contrast-Guided Feature Adjustment Module

Contrastive learning is first proposed to bridge the tasks of OCR and information extraction. The major difficulty is how to design proper representations on behalf of the two tasks. Therefore, we propose CFAM, the key component of our framework, as shown in Fig. 4. Given visual features of each instance $I \in \mathbb{R}^{N \times C}$, the encoded features $E \in \mathbb{R}^{N \times C}$ are obtained via context modeling between I, where N and C indicate the number of instances and the channel of features, respectively. Then E is used to guide the generation of entity features

$L' \in \mathbb{R}^{M \times C}$, where M indicates the number of entity categories. Next, we take E and L' as inputs and calculate a similarity matrix $S \in \mathbb{R}^{N \times M}$ between them. To supervise the generation of S, we use the ground truth $\tilde{S}$ of correspondences between instances and entity categories, which is obtained from the ground truth of the entity. Finally, the outputs of CFAM are S and E.

Generation of Encoded Features. The visual features of each instance I usually are discrete among instances. Thus, we use three Self-Attention (SA) layers to encode $I \in \mathbb{R}^{N \times C}$ and generate the encoded features $E \in \mathbb{R}^{N \times C}$, which can effectively establish connections among instances and transfer rich visual features I from the text detection to the information extraction. The SA consists of three inputs, including queries (Q), keys (K), and values (V), defined as follows:

$$SA(Q, K, V) = softmax(\frac{Q \cdot K^T}{\sqrt{C/m}}) \cdot V, \tag{1}$$

where Q, K, and V are obtained from the same input I (e.g., $Q{=}IW_Q$). Particularly, we use the multi-head self-attention (MSA) to construct the complex feature relations, $MSA = [SA_1; SA_2; ..; SA_m]W_O$, where W_O is a projection matrix and m is the number of attention heads, set as 8.

Generation of Entity Features. It is crucial to design the entity features $L' \in \mathbb{R}^{M \times C}$, which represents the information extraction task. The natural way is directly adopting the learnable entity embeddings $L \in \mathbb{R}^{M \times C}$ as entity features. However, the entity features are almost the same among all samples for the same initialization of learnable entity embeddings and cannot fully use the information from OCR. Additionally, simply using encoded features E as entity features causes lower generalization for information extraction. Thus, based on this consideration, we adopt E and L together to generate the entity features L'. Given the encoded features E, which are first transposed as E^T, then the E^T are fed into FC layer, which results in $E' \in \mathbb{R}^{C \times M}$:

$$E' = FC(E^T) \tag{2}$$

Next, we transform the E', then use E' to guide the generation of the entity features $L' \in \mathbb{R}^{M \times C}$ by adding the E' to learnable entity embeddings $L \in \mathbb{R}^{M \times C}$, defined as follows:

$$L' = L + E'^T \tag{3}$$

Generation of Similarity Matrix. The correspondences between encoded features E and entity features L' are significant for establishing the relations between the tasks of OCR and information extraction. Hence, we take E and L' jointly as inputs to construct the relations between them via calculating the product of the two matrices, which results in a similarity matrix $S \in \mathbb{R}^{N \times M}$ reflecting the relations between encoded features E and entity features L':

$$S = E \cdot L'^T \tag{4}$$

Moreover, we use the ground truth $\tilde{S}$ that is obtained from the annotated labels of the entity to supervise the generation of S.

4.2   Information Extraction Module

Information extraction requires both visual and textual understanding to automatically assign visual elements to various entities. Therefore, the recognition features T, encoded features E, and similarity matrix S, indicating the visual, textual and correlative information are used for information extraction. Following previous representative methods [30,40], we adopt LSTM-CRF [14] as an information extraction module to predict entity categories for all characters at the character level and output the final structure results.

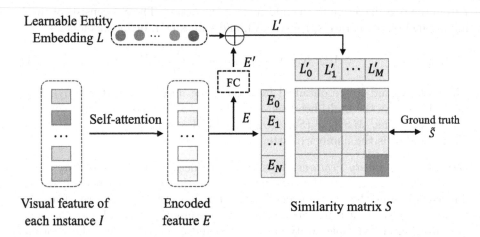

Fig. 4. The structure of the proposed contrast-guided feature adjustment module (CFAM).

4.3   Loss Function

In the training phase, our proposed framework can be trained in an end-to-end manner with the weighted sum of the losses generated from four parts of text detection, recognition, information extraction, and contrastive learning, which are defined as follows:

$$L = \alpha L_{det} + \beta L_{rec} + \gamma L_{ie} + \lambda L_c, \qquad (5)$$

where α, β, γ, and λ are hyper-parameters that control the tradeoff between losses.

L_{det} is the same loss as Mask R-CNN [4] for text detection. L_{rec} is the cross entropy loss for character recognition as follows:

$$L_{rec} = CE(R, \tilde{R}), \qquad (6)$$

where CE indicates the cross entropy loss, $\tilde{R}$ denotes the ground truth of each character.

L_{ie} is also the cross entropy loss for entity classification as follows:

$$L_{ie} = CE(A, \tilde{A}), \tag{7}$$

where CE indicates the cross entropy loss, $\tilde{A}$ denotes the ground truth of entity category.

L_c indicates the loss for contrastive learning. Due to the imbalanced distribution of the number of entities, we adopt the focal loss [19] for L_c as follows:

$$L_c = FL(S, \tilde{S}), \tag{8}$$

where FL indicates the focal loss, $\tilde{S}$ denotes the actual entity category of each character.

Table 2. Performance comparison of the SOTA algorithms on the SROIE and POIE datasets. For text spotting, the results contain two parts (the left and right parts indicate the results of text detection and recognition, respectively). The Δ means the drop of results from SROIE to POIE dataset. * indicates our reproduced results. Note that all results are exhibited in F1-Score.

Setting	Method	SROIE	POIE	Δ
Text Spotting	Mask Textspotter [17]*	97.38/91.23	97.68/91.03	0.30/−0.20
Pure IE	VIES [30]	96.12	84.08*	−12.04
	PICK [38]	96.12	83.23*	−12.89
	TRIE [40]	96.18	82.45*	−13.73
	LayoutLMv2 [33]	96.25	84.18	−12.07
End-to-End IE	TRIE [40]	82.06	76.37*	−5.69
	VIES [30]*	83.44	77.19	−6.25

5 Experiments

5.1 Implementation Details

Our proposed method is implemented in Pytorch. We use four TITAN Xp with 12GB RAM to train our model with batch size four and the Adadelta optimizer. The learning rate starts from 2e-4 and decays to 2e-6 following the linear decay schedule. For the SROIE dataset, the total training epoch is set to 600 for a fair comparison. The total training epoch for the POIE is set to 200 for a fair comparison. We adopt Mask R-CNN [4] as our text detector with ResNet-50 [5] followed by FPN [18]. The hyper-parameters α and β are set to 1.0 while γ and λ are set to 10.0. We choose the best results of all methods in total epoch as the final results. Besides, we use bounding boxes and transcripts for all the text in the images.

5.2   Analysis of Our Proposed Dataset

In this part, to verify the practical utility of our POIE dataset, we make comprehensive comparisons between the POIE dataset and the SROIE [7] dataset. Although SROIE is the most widely used English dataset in the field of VIE, there exists a large number of errors in the annotations of SROIE. Additionally, the SROIE mainly focuses on document images with a single scene layout and few entities, which cannot fully reflect the challenges of information extraction in the wild. The comparisons are from three aspects: 1) We first explore the performance of popular text spotting method [17] on these two datasets. The aim of this setting is to verify the differences between the OCR task and VIE task. 2) Then, we evaluate the performance of a few typical methods for the pure IE task (i.e., directly using the ground truth of bounding boxes and texts as the inputs for information extraction). This setting can effectively reveal the difficulty of variable layouts and entities for information extraction. 3) Finally, to further evaluate the challenges of our dataset, we evaluate some methods in the end-to-end setting (i.e., using OCR results from text spotting as inputs for information extraction.) The details comparisons of these three settings are listed as follows:

Table 3. Performance comparison of the state-of-the-art algorithms on SROIE and POIE datasets. * indicates our reproduced results. Backbone of all algorithms is ResNet50. Note that all methods are evaluated in F1-Score without post-processing.

Method	SROIE	POIE
GRAPHIE [23]	76.51	-
Chargrid [11]	78.24	-
GCN [20]	80.76	-
TRIE [40]	82.06	-
TRIE [40]*	82.82	76.37
VIES [30]*	83.44	77.19
Ours	**85.87**	**79.54**

Text Spotting Setting. As shown in the first part of Table 2, we observe that the performances of the pioneering text spotting algorithm Mask Textspotter [17] are slightly different on POIE and SROIE, which demonstrates that owing to the rapid development of text spotting, document and scene texts can be effectively detected and recognized by the advanced algorithms.

Pure IE Setting. As listed in the second part of Table 2, we find that the performance of representative methods [30,40] have a significant drop from SROIE to our dataset, e.g., the performance of VIES and TRIE from 96.12 to 82.91 and from 96.18 to 81.79 respectively. The reason is that the variable layouts and entities are much more difficult for information extraction, demonstrating that our dataset is more practical for promoting advanced VIE algorithms.

End-to-End IE Setting. As shown in the third part of Table 2, the performance of the TRIE [40] and VIE [30] has distinct differences from SROIE to ours, e.g., the performance of TRIE and VIES from 82.06 to 76.37 and from 83.44 to 77.19, respectively, which further proves our dataset is also more practical for end-to-end methods.

Qualitative Analysis. In this part, we show the qualitative analysis for the difficulty of our POIE dataset. To extract the values of entities from images, the method requires adequate semantic understanding, which is very challenging for algorithms. A typical challenge of our POIE dataset can be seen in Fig. 5. We find that without any clues for the "serving size" (an entity in our dataset), the algorithm cannot accurately predict the value of "serving size".

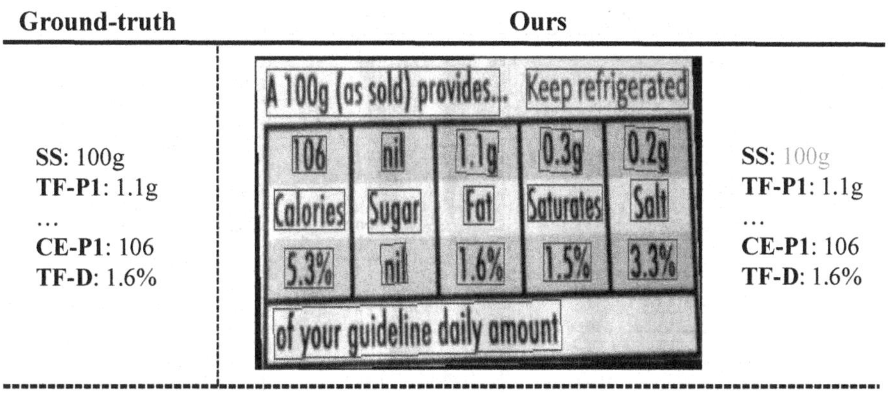

Fig. 5. Qualitative analysis of the proposed POIE dataset. Red boxes indicate the predicted detection results. Texts show the results of information extraction. Note that green texts indicate the false negative results. (Color figure online)

Table 4. Ablation study of CFAM on the proposed POIE dataset. SCH indicates a simple classification head.

SCH	CFAM	POIE dataset		
		F1-Score (DET)	F1-Score (REC)	F1-Score (IE)
-	-	97.50	89.40	75.05
✓	-	97.10	89.01	77.59
-	✓	**97.89**	**91.68**	**79.54**

5.3   The Comparisons of Our Method and SOTA

In this part, we compare our method with previous state-of-the-art (SOTA) methods [11,20,23,30,40] on POIE and SROIE datasets. Our method is in an end-to-end manner, and the core of our approach is to narrow the semantic gap between the tasks of OCR and information extraction. Consequently, we mainly compare our proposed method with the end-to-end methods and focus on the performance of IE (i.e., F1-score of information extraction).

As shown in Table 3, our method achieves SOTA results on POIE and SROIE datasets, and outperforms other methods by an obvious margin. For the POIE dataset, our method exhibits superior performance. For the SROIE dataset, most previous methods do not use post-progress, thus we mainly focus on the results produced without post-progress. Additionally, our method outperforms other methods by obvious gains. Moreover, Fig. 6 shows some qualitative results of ours and other methods. We can observe that our method is more precisely to predict the values of entities.

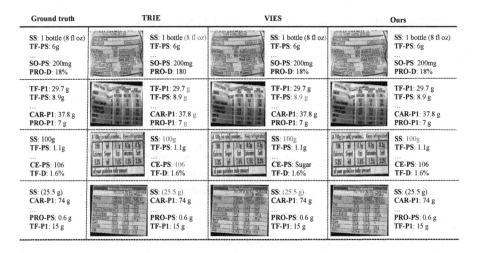

Fig. 6. Some visualization results of our method and others on the proposed POIE dataset. There are results of TRIE, VIES, and ours, from left to right, respectively. Red boxes indicate the predicted detection results. Texts show the information extraction results. Note that green and blue indicate the false negative and false positive results, respectively. (Color figure online)

5.4   Ablation Studies

The ablation studies are carried out on our POIE dataset. For a complete comparison, we show the experimental results of DET, REC, and IE to evaluate the effectiveness of our proposed modules. Specifically, as described below:

Effectiveness of CFAM. To verify the effectiveness of CFAM, we conduct experiments in the settings: classification and CFAM. Classification indicates directly classifying visual features to the corresponding entity. As listed in Table 4, we observe that classification can promote the performance of information extraction but have a negative impact on text detection and recognition tasks. We argue that it is because our method is an end-to-end method, where the OCR and information extraction tasks are joint optimizations. Moreover, compared with classification, we find that with CFAM, our method achieves superior results (97.89% in DET, 91.68% in REC, 79.54% in IE) for the reason that CFAM serves as an auxiliary module to bridge the relations between the tasks of OCR and information extraction. Additionally, to verify the necessity of using supervision for contrastive learning in CFAM, we conduct experiments in the setting: with L_c and without L_c. As shown in Table 5, we found that the contrastive loss L_c can promote the performance.

Analysis of Generated Entity Features. In our proposed CFAM module, it is important to design the entity features representing the information extraction task. There are three strategies for generating the entity features (i.e., directly using learnable entity embeddings, adopting encoded features, and combining the learnable entity embeddings and encoded features). The performance of these strategies is shown in Table 6. We observe that using learnable entity embeddings and encoded features together achieves the best results. However, either using learnable entity embeddings or encoded features will get lower results. We argue that the reasons are two-fold: 1) without encoded features, the generated entity features are almost the same among all entity features and cannot fully use the information from instance. 2) without learnable entity embeddings, the generated entity features completely rely on the encoded feature and cannot fully represent the feature of information extraction.

Table 5. The effectiveness of L_c in CFAM.

L_c	POIE dataset		
	F1-Score (DET)	F1-Score (REC)	F1-Score (IE)
-	97.48	91.54	78.69
✓	**97.89**	**91.68**	**79.54**

Table 6. Analysis of generated entity features. Note that all results are shown in F1-Score.

Learnable entity embeddings	Encoded features	POIE dataset		
		DET	REC	IE
✓	-	97.48	90.51	78.86
-	✓	97.81	91.48	76.25
✓	✓	**97.89**	**91.68**	**79.54**

Generalization of CFAM. To further demonstrate the generalization of our proposed CFAM, we evaluate CFAM by applying it to other SOTA methods [30,40].

As listed in Table 7, the proposed CFAM improves the performance of TRIE and VIES by a distinguish margin even though they are strong methods. For our POIE dataset, CFAM improves the SOTA method TRIE and VIES by 2.45% and 1.92%, respectively. For the SROIE dataset, CFAM enhances the performance of TRIE and VIES by an obvious margin. The results demonstrate the generalization of our proposed CFAM, which can be effectively applied to other methods.

5.5 Limitation

Our method has achieved superior results on SROIE and ours. However, we find that the challenges, e.g., different forms for the same entity (Fig. 1(b)), have not been solved well and still influence the performance of information extraction. In the future, we will further explore the commonality between different forms of the same entity and design a new method to effectively solve this problem.

Table 7. The effectiveness of CFAM in other methods on the SROIE and POIE datasets. We reproduce the TRIE and VIES. Note that all results are exhibited in F1-Score.

Method	SROIE	POIE
TRIE [40]	82.82	76.37
TRIE [40] + CFAM (**ours**)	**84.25**	**79.17**
VIES [30]	83.44	77.19
VIES [30] + CFAM (**ours**)	**84.32**	**79.11**

6 Conclusion

In this paper, we have proposed a large-scale English dataset (called POIE) consisting of camera images, which can reflect the challenges of the real world. We have designed a novel end-to-end framework with a plug-and-play CFAM for VIE tasks, which adopts contrastive learning and properly designs the representation of VIE tasks for contrastive learning. The experimental results prove that our dataset is more practical for promoting advanced VIE algorithms. Additionally, the experiments demonstrate that our proposed framework consistently achieves the obvious performance gains on SROIE and ours. In the future, we will consider to extend the dataset with more images, entities, and diverse natural scene disturbances. We hope our proposed dataset and framework can promote further investigation in VIE.

Acknowledgements. This work was supported by the National Science Fund for Distinguished Young Scholars of China (Grant No. 62225603).

References

1. Aberdam, A., et al.: Sequence-to-sequence contrastive learning for text recognition. In: Proceedings of the IEEE/CVF Conference on Computer Vision and Pattern Recognition, pp. 15302–15312 (2021)
2. Cao, H., et al.: Query-driven generative network for document information extraction in the wild. In: ACM MM, pp. 4261–4271 (2022)
3. Guo, H., Qin, X., Liu, J., Han, J., Liu, J., Ding, E.: Eaten: entity-aware attention for single shot visual text extraction. In: 2019 International Conference on Document Analysis and Recognition (ICDAR), pp. 254–259. IEEE (2019)
4. He, K., Gkioxari, G., Dollár, P., Girshick, R.: Mask R-CNN. In: ICCV, pp. 2961–2969 (2017)
5. He, K., Zhang, X., Ren, S., Sun, J.: Deep residual learning for image recognition. In: Proceedings of the IEEE Conference on Computer Vision and Pattern Recognition, pp. 770–778 (2016)
6. Huang, J., Guan, D., Xiao, A., Lu, S.: Model adaptation: historical contrastive learning for unsupervised domain adaptation without source data. Adv. Neural. Inf. Process. Syst. **34**, 3635–3649 (2021)
7. Huang, Z., et al.: ICDAR 2019 competition on scanned receipt OCR and information extraction. In: 2019 International Conference on Document Analysis and Recognition (ICDAR), pp. 1516–1520. IEEE (2019)
8. Hwang, W., et al.: Post-OCR parsing: building simple and robust parser via bio tagging (2019)
9. Jiang, X., Long, R., Xue, N., Yang, Z., Yao, C., Xia, G.S.: Revisiting document image dewarping by grid regularization. In: Proceedings of the IEEE/CVF Conference on Computer Vision and Pattern Recognition, pp. 4543–4552 (2022)
10. Kahraman, H.T., Sagiroglu, S., Colak, I.: Development of adaptive and intelligent web-based educational systems. In: 2010 4th International Conference on Application of Information and Communication Technologies, pp. 1–5. IEEE (2010)
11. Katti, A.R., et al.: Chargrid: towards understanding 2D documents. In: EMNLP (2018)
12. Khosla, P., et al.: Supervised contrastive learning, vol. 33, pp. 18661–18673 (2020)
13. Kim, G., et al.: OCR-free document understanding transformer. In: Avidan, S., Brostow, G., Cissé, M., Farinella, G.M., Hassner, T. (eds.) ECCV 2022. LNCS, vol. 13688, pp. 498–517. Springer, Cham (2022). https://doi.org/10.1007/978-3-031-19815-1_29
14. Lample, G., Ballesteros, M., Subramanian, S., Kawakami, K., Dyer, C.: Neural architectures for named entity recognition. In: Proceedings of NAACL-HLT, pp. 260–270 (2016)
15. Li, W., et al.: Unimo: towards unified-modal understanding and generation via cross-modal contrastive learning. In: Proceedings of the 59th Annual Meeting of the Association for Computational Linguistics and the 11th International Joint Conference on Natural Language Processing (Volume 1: Long Papers), pp. 2592–2607 (2021)
16. Li, X.H., Yin, F., Dai, H.S., Liu, C.L.: Table structure recognition and form parsing by end-to-end object detection and relation parsing. Pattern Recogn. **132**, 108946 (2022)
17. Liao, M., Pang, G., Huang, J., Hassner, T., Bai, X.: Mask TextSpotter v3: segmentation proposal network for robust scene text spotting. In: Vedaldi, A., Bischof, H., Brox, T., Frahm, J.-M. (eds.) ECCV 2020. LNCS, vol. 12356, pp. 706–722. Springer, Cham (2020). https://doi.org/10.1007/978-3-030-58621-8_41

18. Lin, T.Y., Dollár, P., Girshick, R., He, K., Hariharan, B., Belongie, S.: Feature pyramid networks for object detection. In: Proceedings of the IEEE Conference on Computer Vision and Pattern Recognition, pp. 2117–2125 (2017)
19. Lin, T.Y., Goyal, P., Girshick, R., He, K., Dollár, P.: Focal loss for dense object detection. In: ICCV, pp. 2980–2988 (2017)
20. Liu, X., Gao, F., Zhang, Q., Zhao, H.: Graph convolution for multimodal information extraction from visually rich documents. In: Proceedings of the 2019 Conference of the North American Chapter of the Association for Computational Linguistics: Human Language Technologies, Volume 2 (Industry Papers), pp. 32–39 (2019)
21. Park, S., et al.: Cord: a consolidated receipt dataset for post-OCR parsing (2019)
22. Patashnik, O., Wu, Z., Shechtman, E., Cohen-Or, D., Lischinski, D.: Styleclip: text-driven manipulation of stylegan imagery. In: Proceedings of the IEEE/CVF International Conference on Computer Vision, pp. 2085–2094 (2021)
23. Qian, Y., Santus, E., Jin, Z., Guo, J., Barzilay, R.: Graphie: a graph-based framework for information extraction. In: Proceedings of the 2019 Conference of the North American Chapter of the Association for Computational Linguistics: Human Language Technologies, Volume 1 (Long and Short Papers), pp. 751–761 (2019)
24. Radford, A., et al.: Learning transferable visual models from natural language supervision. In: International Conference on Machine Learning, pp. 8748–8763. PMLR (2021)
25. Rao, Y., et al.: Denseclip: language-guided dense prediction with context-aware prompting. In: Proceedings of the IEEE/CVF Conference on Computer Vision and Pattern Recognition, pp. 18082–18091 (2022)
26. Sun, H., Kuang, Z., Yue, X., Lin, C., Zhang, W.: Spatial dual-modality graph reasoning for key information extraction. arXiv preprint arXiv:2103.14470 (2021)
27. Tang, G., et al.: Matchvie: exploiting match relevancy between entities for visual information extraction (2021)
28. Wang, H., Bai, X., Yang, M., Zhu, S., Wang, J., Liu, W.: Scene text retrieval via joint text detection and similarity learning. In: Proceedings of the IEEE/CVF Conference on Computer Vision and Pattern Recognition, pp. 4558–4567 (2021)
29. Wang, H., et al.: Knowledge mining with scene text for fine-grained recognition. In: Proceedings of the IEEE/CVF Conference on Computer Vision and Pattern Recognition, pp. 4624–4633 (2022)
30. Wang, J., et al.: Towards robust visual information extraction in real world: new dataset and novel solution. In: Proceedings of the AAAI Conference on Artificial Intelligence, vol. 35, pp. 2738–2745 (2021)
31. Wang, W., et al.: mmlayout: multi-grained multimodal transformer for document understanding. In: Proceedings of the 30th ACM International Conference on Multimedia, pp. 4877–4886 (2022)
32. Wong, K.Y., Casey, R.G., Wahl, F.M.: Document analysis system. IBM J. Res. Dev. 26(6), 647–656 (1982)
33. Xu, Y., et al.: Layoutlmv2: multi-modal pre-training for visually-rich document understanding. In: Proceedings of the 59th Annual Meeting of the Association for Computational Linguistics and the 11th International Joint Conference on Natural Language Processing (Volume 1: Long Papers), pp. 2579–2591 (2021)
34. Xu, Y., Li, M., Cui, L., Huang, S., Wei, F., Zhou, M.: Layoutlm: pre-training of text and layout for document image understanding. In: ACMSIGKDD, pp. 1192–1200 (2020)

35. Yang, M., et al.: Reading and writing: Discriminative and generative modeling for self-supervised text recognition. In: Proceedings of the 30th ACM International Conference on Multimedia, pp. 4214–4223 (2022)
36. You, Y., Chen, T., Sui, Y., Chen, T., Wang, Z., Shen, Y.: Graph contrastive learning with augmentations, vol. 33, pp. 5812–5823 (2020)
37. Yu, W., Liu, Y., Hua, W., Jiang, D., Ren, B., Bai, X.: Turning a clip model into a scene text detector. arXiv preprint arXiv:2302.14338 (2023)
38. Yu, W., Lu, N., Qi, X., Gong, P., Xiao, R.: Pick: processing key information extraction from documents using improved graph learning-convolutional networks. In: ICPR, pp. 4363–4370. IEEE (2021)
39. Zeng, G., et al.: Beyond OCR+ VQA: towards end-to-end reading and reasoning for robust and accurate TextVQA. Pattern Recognit. **138**, 109337 (2023)
40. Zhang, P., et al.: TRIE: end-to-end text reading and information extraction for document understanding. In: Proceedings of the 28th ACM International Conference on Multimedia, pp. 1413–1422 (2020)
41. Zhou, K., Yang, J., Loy, C.C., Liu, Z.: Learning to prompt for vision-language models. IJCV **130**(9), 2337–2348 (2022)
42. Zhu, F., Lei, W., Feng, F., Wang, C., Zhang, H., Chua, T.S.: Towards complex document understanding by discrete reasoning. In: Proceedings of the 30th ACM International Conference on Multimedia, pp. 4857–4866 (2022)

Scene Text Recognition with Image-Text Matching-Guided Dictionary

Jiajun Wei[1], Hongjian Zhan[1,2], Xiao Tu[1], Yue Lu[1(✉)], and Umapada Pal[3]

[1] Shanghai Key Laboratory of Multidimensional Information Processing, East China Normal University, Shanghai, China
jjwei@stu.ecnu.edu.cn, ecnuhjzhan@foxmail.com, xtu@cee.ecnu.edu.cn, ylu@cs.ecnu.edu.cn
[2] Chongqing Institute of East China Normal University, Chongqing 401120, China
[3] CVPR Unit, Indian Statistical Institute, Kolkata, India
umapada@isical.ac.in

Abstract. Employing a dictionary can efficiently rectify the deviation between the visual prediction and the ground truth in scene text recognition methods. However, the independence of the dictionary on the visual features may lead to incorrect rectification of accurate visual predictions. In this paper, we propose a new dictionary language model leveraging the **S**cene **I**mage-**T**ext **M**atching(SITM) network, which avoids the drawbacks of the explicit dictionary language model: 1) the independence of the visual features; 2) noisy choice in candidates etc. The SITM network accomplishes this by using Image-Text Contrastive (ITC) Learning to match an image with its corresponding text among candidates in the inference stage. ITC is widely used in vision-language learning to pull the positive image-text pair closer in feature space. Inspired by ITC, the SITM network combines the visual features and the text features of all candidates to identify the candidate with the minimum distance in the feature space. Our lexicon method achieves better results(93.8% accuracy) than the ordinary method results(92.1% accuracy) on six mainstream benchmarks. Additionally, we integrate our method with ABINet and establish new state-of-the-art results on several benchmarks.

Keywords: Dictionary Language Model · Scene Image-Text Matching · Image-Text Contrastive Learning · Scene Text Recognition

1 Introduction

Deep learning-based scene text recognition has been developed for years. The accuracy of scene text recognition has vastly increased as the appropriate design of model architecture and the expansion of model size. Previous methods [4,5, 36,48] can address a variety of recognition issues, but the inherent ambiguities, such as complicated background or diversity of font, etc, render the recognized results inaccurate.

G. A. Fink et al. (Eds.): ICDAR 2023, LNCS 14192, pp. 54–69, 2023.
https://doi.org/10.1007/978-3-031-41731-3_4

Due to the unique characteristics of text recognition, it is feasible to employ human language priors to rectify the output of a vision recognition model. Utilizing a pre-trained language model is one of the common methods. Fang et al. [7] pre-train a language model using WikiText-103 [28]. The pre-trained language model rectifies the visual prediction through learning grammar and the construction of words in the human language system. Another popular approach is to search for a word that has minimum edit distance(Levenshtein distance [19]) with the visual prediction in a dictionary. Nguyen et al. [31] present a method for incorporating a dictionary into the training pipeline. They use the dictionary to generate a certain number of candidates and then output the most compatible one with the highest compatibility scores in a probability matrix $\mathbf{P}$, which is generated by the visual feature $\boldsymbol{F}_v$. But they still disregard the interaction between visual features and text features in the inference stage.

The aforementioned methods utilizing explicit language models have several problems. First, regardless of the pre-trained language model or dictionary language model, the independence of the language model from the visual feature may erroneously rectify the correct prediction results. Second, it is illogical to utilize human language priors to rectify texts that appear infrequently or have no linguistic information(*e.g.* ngee, tsc), since neither a pre-trained language model nor a dictionary can rectify texts without human language logic.

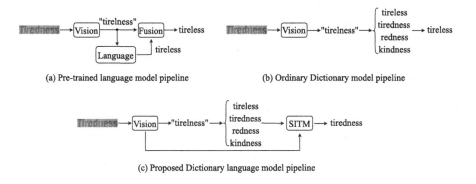

(a) Pre-trained language model pipeline   (b) Ordinary Dictionary model pipeline

(c) Proposed Dictionary language model pipeline

Fig. 1. Comparison with different pipelines.

In this paper, we present an effective method for incorporating a dictionary into scene text recognition that possesses two advantages: 1) taking visual prediction into account. When generating candidates for the visual prediction, the visual prediction is also included in the candidate set. In this case, the initial visual prediction has a probability to be the ultimate outcome. 2) Integrating visual features into the inference stage. Image-Text Contrastive(ITC) Learning is an unsupervised learning method that aims to make positive image-text pairs have higher similarity scores. Inspired by ITC learning, we additionally integrate a **S**cene **I**mage-**T**ext **M**atching(SITM) network to match visual features and text features by ITC Learning. Nevertheless, when we merely employ other label texts in the same batch as negatives, the image-text matching accuracy in the inference stage is not as good as the training stage. We address the problem by generating hard negatives that resemble the shape of label texts. The difference between three methods is depicted in Fig. 1

The main contributions of this paper are summarized as follows: 1) we propose a novel method to integrate a dictionary into scene text recognition that avoids the drawbacks of an ordinary dictionary language model. 2) We also offer a new strategy that employs labels to generate resemblant words as hard negatives in the SITM training stage. 3) A Scene Image-Text Matching Module is introduced, which matches positive image-text pairs in the inference stage.

2   Related Work

2.1   Scene Text Recognition

Language-Free Methods. Language-free approaches typically provide a prediction based on visual features, regardless of context information. CTC-based methods [8] utilize CNN to extract visual features, RNN to model sequence features, and CTC loss to train the entire recognition network end-to-end [10,11,38]. Segmentation-based methods [22] segment each character region before classifying and recognizing. The recognition results of all character areas compose the entire text sequence [27,41,46]. However, due to the absence of context information interaction, these approaches cannot attain exceptional performance.

Language-Based Methods. In previous works, [13,14] use explicit language models to improve model recognition accuracy. CNN is employed to extract visual features to predict bags of N-grams of text strings. Recently, [7] regards the explicit language model as a spell checker to rectify visual prediction results. Some implicit language-based approaches connect visual features with context information by utilizing RNN [18,37] or attention mechanisms [35,43]. First, an image encoder is employed to extract features from word images, follower by an attention-based method for integrating visual features and context information. [4–6,36] focus on relevant information from 1D image features, and [20,23,45,48] from 2D image features. Some performance-enhancing approaches focus on learning new feature representations. [1,26,47] train their models by sequence contrastive learning, masked image modeling, and a mix of the two, respectively.

2.2   Vision-Language Learning

There are two categories of visual-language representation learning. In the first category, text features and image features are fused using a multi-mode encoder [21,24,25,39]. This type of approaches has achieved outperformance in downstream tasks such as NLVR [40] and VQA [2]. The Second category focus on learning separate texts and images encoders [15,33]. CLIP [33] employs contrastive loss to train the image encoder and text encoder on a massive quantity of network image-text pairs. We opt for the second category to reuse the visual encoder trained in the recognition stage instead of training a new visual encoder. Then, the text encoder is trained from scratch in the matching stage.

3   Method

We propose a new method to incorporate a dictionary into scene text recognition. The dictionary is used to generate the certain number of candidates, which will subsequently be matched with the visual features by SITM to output the candidate with the highest similarity score. In this section, the details of the overall architecture are presented. We will also describe the Resemblant word generation strategy and the SITM network. The objective training function is finally introduced.

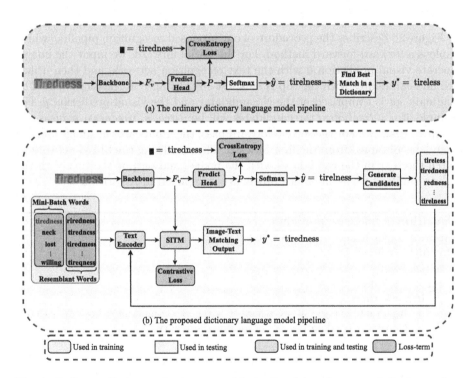

Fig. 2. Ordinary dictionary language model pipeline (a) and proposed dictionary language model pipeline (b). In the ordinary pipeline, the prediction is forced to be one with the smallest edit distance in the dictionary. In the proposed pipeline, the ultimate prediction is determined by SITM

3.1 Overall Architecture

As can be seen in Fig. 3, a general scene text recognition framework usually consists of a feature extraction module, a sequence modeling module, and a prediction module, which was proposed by Baek et al. [3]. Our proposed dictionary method can combine any scene text recognition method with the above framework. We utilize the output of the sequence modeling as the visual features F_v. Specifically in this paper, we employ the vision module of the ABINet [7] as our baseline network.

Fig. 3. General Vision Architecture of Scene Text Recognition.

Figure 2b describes the procedure of our proposed recognition pipeline, which employs a forward-forward method. For the initial forward, we input the image, generate visual prediction $\hat{y}$ with the text recognition network, and then utilize $\hat{y}$ to find candidates with the top N smallest edit distance in the dictionary. The candidate set is comprised of the N candidates and the visual prediction $\hat{y}$. For example, if $\hat{y} = tirelness$, the candidates will be: *tireless, tiredness, redness, ..., kindness, tirelness*. For the second forward pass, the inputs are the image and the candidates obtained from the first forward. Then, utilizing the SITM network to match the text in the candidates with the image, and output the text with the highest similarity. The inference procedure is depicted in Algorithm 1.

Algorithm 1 Inference procedure

Input: x: Input Image; n: Forward State
Output: Prediction y^*

1: initial $n = 1$
2: **for** $n = 1, 2$ **do**
3:     **if** $n = 1$ **then**
4:         Input x to get $\hat{y} = \mathbf{V}(x)$, where $\mathbf{V}$ is the vision module
5:         Construct candidate set C of $\hat{y}$ using dictionary
6:     **else**
7:         Input x and C to calculate the similarity scores S
8:         Get the $c \in C$ with the highest score in S as the final prediction y^*
9: **return** y^*

During training, we generate text candidates for Image-Text Contrastive Learning by using labels in the same mini-batch. In addition, we create a certain number of resemblant words as hard negatives. A contrastive loss function, which is defined based on the similarity cross-entropy function depicted in Sect. 3.3, and the recognition loss are then employed for training.

3.2 Resemblant Words Generation

Image	Label	Hard Dictionary Negatives
	tiredness	'redness' 'tireless' 'kindness' 'likeness' 'redress' 'timeless' 'nakedness'
	short	'sort' 'shot' 'shorts' 'shirt' 'sport' 'snort' 'shout'
	break	'bleak' 'bread' 'creak' 'freak' 'wreak' 'beak' 'beau'
	could	'cold' 'would' 'bold' 'gourd' 'cod' 'should' 'court'

Fig. 4. Qualitative hard negatives in the inference stage.

There is a gap between the SITM training stage and the inference stage if we utilize the normal contrastive learning method. Specifically, in the training stage, the negatives are the other labels in the same mini-batch for a single image. For example, if a mini-batch contains: *tiredness, kills, short, break, could, save, **your**, life*, the text negatives are *tiredness, kills, short, break, could, save, life* and the text positive is ***your*** for the image ***your***. However, in the inference stage, the negatives are candidates from the dictionary with the top N smallest edit distance, which is similar to the ground truth. For example, for the image ***your***, the negatives in the inference stage are *pour, you, tour, hour, dour, sour, four*. Figure 4 exhibits some hard negatives for the labels in the test set. We find the gap would cause some mismatches between image-text pairs in the inference stage and degrade the performance of the dictionary.

We address the problem with our proposed Resemblant Words Generation strategy. When we train the SITM network, in addition to using the text labels corresponding to other images in the same batch as negatives, we present a strategy for constructing hard negatives using labels. Specifically, we initially establish a similar character lookup table containing five similar characters for each English character. We observe the difference between visual prediction and ground truth and record the wrong predicted characters as the composition of the lookup table. For character a, we select d, e, o, q and u as the similar characters. Then we randomly replace a character in a label having a similar appearance. For example, if y = *tiredness* and the number of the resemblant words is 4, the hard negatives will be *riredness, tiredncss, tiredmess, tireqness*.

3.3 Scene Image-Text Matching Module

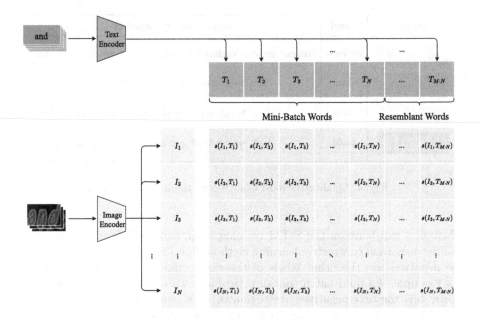

Fig. 5. Scene Image-Text Matching Module Architecture.

The Scene Image-Text Matching module contains image encoder, text encoder and Scene Image-Text Contrastive Learning module. The image encoder consists of a backbone network that shares parameters with the backbone network of the recognition module and a parallel attention layer that is used to convert visual features to sequence features. The text encoder consists of two layers of transformer encoder. Scene Image-Text Contrastive Learning module consists of two liner layers. Image sequence features $I \in \mathbb{R}^{L \times C}$ and text sequence features $T \in \mathbb{R}^{L \times C}$ are obtained by image encoder and text encoder, respectively. The image features I and text features T pass through the linear projection layer before Image-Text Contrastive Learning. Figure 5 shows the details of our SITM module.

During the training stage, we employ the image-text contrastive learning task in vision-language learning to complete the scene image-text matching task. Image-Text Contrastive Learning aims to learn a representation of distinct modal features. It learns the cosine similarity function $s(\boldsymbol{I}, \boldsymbol{T}) = \frac{l_v(\boldsymbol{I}) \cdot l_t(\boldsymbol{T})}{||l_v(\boldsymbol{I})|| \cdot ||l_t(\boldsymbol{T})||}$, where l represents linear layer and $\boldsymbol{I} \in \mathbb{R}^{L \times C}$, $\boldsymbol{T} \in \mathbb{R}^{L \times C}$ represent image features and text features, respectively. The matched image-text pair will have a higher similarity score. We calculate image-to-text(i2t) and text-to-image(t2i) similarity and normalize the results using softmax. The formulas are as below:

$$p_m^{i2t}(I) = \frac{\exp(s(I,T_m)/\tau)}{\sum_{n=1}^{MN}\exp(s(I,T_n)/\tau)}, \quad p_m^{t2i}(T) = \frac{\exp(s(T,I_m)/\tau)}{\sum_{n=1}^{N}\exp(s(T,I_n)/\tau)}, \quad (1)$$

where τ is a temperature parameter, m is the order indication of the image or text, s is the cosine similarity function and N is the batch size and M-1 is the number of resemblant words of one label. As can be seen in Fig. 6, the parallel attention layer focuses on the main character features in the image to guide the matching procedure.

Fig. 6. An example of the parallel attention layer Gradient-weighted Class Activation Mapping. The left image is the input and the right image is the activation mapping.

During the inference stage, we only calculate the image-to-text(i2t) similarity to find the highest score among candidates from the candidate set:

$$p_m^{i2t}(I) = \frac{\exp(s(I,T_m)/\tau)}{\sum_{n=1}^{T}\exp(s(I,T_n)/\tau)}, \quad (2)$$

where T is the number of candidates in the candidates set which is depicted in Sect. 3.1.

3.4 Overall Objective Function

A supervised cross-entropy loss function is utilized for training and optimizing the scene text recognizer. By minimizing the negative log-likelihood sequence probability loss function, the difference between the prediction and the ground truth is quantified. The specific formula is given below:

$$\mathcal{L}_{recog} = \boldsymbol{E}_{(x,y)\sim(X,Y)}\{-\mathrm{log}p(y|F_v(x))\}. \quad (3)$$

The full loss function of the Scene Image-Text Matching module consists of $\mathcal{L}_{itc}$

$$\mathcal{L}_{SITM} = \mathcal{L}_{itc}, \quad (4)$$

$$\mathcal{L}_{itc} = \frac{1}{2}\boldsymbol{E}_{(I,T)\sim D}\big[\mathrm{H}(\boldsymbol{y}^{i2t}(I),\boldsymbol{p}^{i2t}(I)) + \mathrm{H}(\boldsymbol{y}^{t2i}(T),\boldsymbol{p}^{t2i}(T))\big], \quad (5)$$

where $\boldsymbol{y}^{i2t}(I)$ and $\boldsymbol{y}^{t2i}(T)$ denote the ground-truth one-hot similarity, in which the negative pair probability is 0 and the positive pair probability is 1. The H is defined as the cross-entropy loss.

The overall objective function $\mathcal{L}_{overall}$ is defined as:

$$\mathcal{L}_{overall} = \lambda_1 \mathcal{L}_{recog} + \lambda_2 \mathcal{L}_{SITM}, \tag{6}$$

where λ_1 and λ_2 are the hyper-parameters used to control the training stages. We respectively set $\lambda_1 = 1, \lambda_2 = 0$ and $\lambda_1 = 0, \lambda_2 = 1$ when we train the recognition module and Scene Image Text Matching module.

4   Experimental Results

4.1   Datasets

The common synthetic datasets SynthText [9] and MJSynth [12] are utilized to train our proposed model. We employ six widely used benchmarks to evaluate the performance of the model, including three regular text datasets ICDAR2013, SVT, IIIT5K and three irregular text datasets ICDAR2015, SVTP and CUTE80. Following are the specifics of the datasets:

ICDAR2013(IC13) [17] has 1015 test images. The dataset contains only horizontal text instances.

Street View Text (SVT) [42] contains 647 images collected from Google Street View. This dataset contains fuzzy, blurry, and low-resolution text images.

IIIT5K [29] contains 3000 test images crawled from Google image searches with query words. Most text instances are rules for horizontal layout.

ICDAR2015(IC15) [16] contains 1811 test images created for the ICDAR 2015 Robust Reading competitions. Most instances of text are irregular (noisy, blurry, perspective or curved).

Street View Text Perspective (SVTP) [32] contains 645 cropped images from Google Street View. Many of the images have a distorted perspective.

CUTE80 [34] is collected from nature scenes and contains 288 cropped images for verification. Most of them are curved text.

4.2   Training Setting

PyTorch is applied to implement the model proposed in this paper. All the experiments are conducted on a 24GB-memory NVIDIA3090. All input images are scaled to 32×128 while maintaining their aspect ratio.The character set includes 37 classes, which contains 10 digits, 26 lowercase letters, and an EOS token. The maximum sequence length is 25. Adam is selected as the optimizer, and the batch size is set to 320.

The training procedure consists of two stages: the recognition stage and the matching stage. In the recognition stage, the text recognition network is merely trained to minimize the text recognition loss function. We trained the recognition network 8 epochs on SynthText and MJSynth from scratch. During the matching stage, the SITM network is unfrozen. Image-Text contrastive loss is applied to train the text encoder.

4.3 Comparison with the Ordinary Dictionary Method and the State-of-the-art

In this part, we compare the accuracy of the baseline, ordinary dictionary-guided baseline and the proposed dictionary-guided baseline on the six benchmarks. The baseline described in Sect. 3.1 serves as a comparison standard in our experiments. We utilize the same lexicon, which comprises approximately 20K words and is composed of numbers, common English words and common English trademarks. *Full Lexicon* is not utilized to construct the dictionary, which means that some words in the test set may not be in the lexicon. As we consider this would be a more realistic dictionary composition with some words included and some were excluded. For a fair comparison, all the methods are trained on the SynthText and MJSynth datasets.

Table 1. Comparison with Ordinary Dictionary-guided Baseline.

Method	Regular Text			Irregular Text			Average
	IC13	SVT	IIIT5K	IC15	SVTP	CUTE80	
Baseline	94.9	90.4	94.6	81.7	84.2	86.5	89.8
Baseline+Dict Guided	95.8	92.1	96.2	85.6	87.4	90.8	92.1
Baseline+Our Method	**97.8**	**94.1**	**97.1**	**88.0**	**89.3**	**93.4**	**93.8**
Improvement	**+2.0**	**+2.0**	**+0.9**	**+2.4**	**+1.9**	**+2.6**	**+1.7**

As can be seen from Table 1, utilizing ordinary dictionary guidance would enhance performance, but the improvement on some benchmarks is insignificant. On six benchmarks, our proposed dictionary-guided method outperforms the ordinary method with 2.0%, 2.0%, 0.9%, 2.4%, 1.9% and 2.6% on IC13, SVT, IIIT5K, IC15, SVTP and CUTE80 datasets, respectively. We also discover that our method has superiority on irregular datasets IC15, SVTP and CUTE80 as they contain low-quality images such as curved and blurred texts. As the visual prediction of the irregular datasets often have more severe deviation from the ground truth, the candidate with smallest edit distance may not be the correct answer.

The weaker performance of the ordinary dictionary method stems from two aspects. 1) It disregards visual prediction. The ordinary dictionary pipeline takes the word in the dictionary as output with the smallest edit distance for the visual prediction that is not a component of the dictionary, which makes the correct prediction incorrect. 2) For words in the dictionary with the same edit distance, the traditional dictionary pipeline is unable to determine which output is right. The random selection will fail to output the correct outcome among the candidates.

Our proposed pipeline effectively avoids the aforementioned issues. In addition to the text recognition network, we also train a SITM network. When a dictionary is employed, we combine the prediction and the top N smallest edit distance dictionary words as candidates set, and the SITM network is used to

Input				
Label	ronaldo	ebizu	bud	finest
Baseline	ronaldo	ebizu	bod	vinest
Ordinary Dict	renal	biz	bold	vines
Proposed Dict	ronaldo	ebizu	bud	finest

Fig. 7. Qualitative results for the ordinary dictionary method and our proposed method.

determine which one is correct. Figure 7 illustrates instances successfully recognized by our method while ordinary dictionary method could not. The second and third columns represent that the visual prediction of the scene text recognition network is correct, but there is no corresponding in the dictionary. In this case, the ordinary method generates the wrong answer. However, our proposed method can find the visual prediction output in candidates. The fourth and fifth columns represents the deviation between the visual predictions and the ground truths. When facing candidates with the same edit distance, the ordinary method can only randomly output, while our proposed method can find the correct candidate word according to the SITM network.

Table 2. Comparison with State-of-the-art Methods and Ordinary Dictionary-guided State-of-the-art Methods.

Methods	Ordinary Dict Guide	Proposed Dict Guide	Regular Text			Irregular Text		
			IC13	SVT	IIIT5K	IC15	SVTP	CUTE80
PlugNet [30]	-	-	95.0	92.3	94.4	82.2	84.3	85.0
SRN [49]	-	-	95.5	91.5	94.8	82.7	85.1	87.8
RobustScanner [50]	-	-	94.1	89.3	95.4	79.2	82.9	92.4
TextScanner [41]	-	-	94.9	92.7	95.7	83.5	84.8	91.6
AutoSTR [51]	-	-	94.2	90.9	94.7	81.8	81.7	-
VisionLAN [44]	-	-	95.7	91.7	95.8	83.7	86.0	88.5
CRNN [3]	-	-	88.8	78.9	84.3	61.5	64.8	61.3
ABINet [7]	-	-	**97.4**	93.5	96.2	86.0	**89.3**	89.2
PARSeq [4]	-	-	97.0	**93.6**	**97.0**	**86.5**	88.9	**92.2**
CRNN [3]	✓		95.2	90.8	91.5	83.0	84.0	78.5
CRNN [3]	-	✓	**96.9**	**92.1**	**93.2**	**84.6**	**88.5**	**80.9**
ABINet [7]	✓		97.7	94.1	96.8	87.5	90.0	90.3
ABINet [7]	-	✓	**98.4**	**95.8**	**98.0**	**88.6**	**90.1**	**91.3**

To verify the effectiveness of our method, we combine two existing scene text recognition frameworks with our proposed dictionary method. We select the CRNN and the state-of-the-art ABINet to validate our proposed approach.

Table 2 shows that our proposed method still outperforms the ordinary dictionary method. As can be seen from the comparison, in the CRNN [3], our proposed dictionary-guided method outperforms the ordinary method with 1.7%, 1.3%, 1.7%, 1.6%, 4.5% and 2.4% on IC13, SVT, IIIT5K, IC15, SVTP, CUTE80 datasets, respectively. In ABINet [7], the improvements on the six benchmarks are 0.7%, 1.7%, 1.2%, 1.1%, 0.1% and 1.0%, respectively. In the meanwhile, we find that the utilization of a dictionary to rectify the visual prediction is a highly effective way of enhancing performance. When employing a dictionary to rectify visual prediction, the CRNN [3] exceeds numerous state-of-the-art methods on some benchmarks.

4.4 Ablation Study

Table 3. Comparison of recognition accuracy on different numbers of candidates.

candidates	1	5	10	20	30	80	150	300
Recognition Accuracy	92.1	93.8	94.1	94.2	94.2	94.3	94.3	94.3

The Recognition Accuracy of Baseline as the Numbers of Candidates Varies: The quantity of candidates is one of the primary distinctions between our approach and the ordinary pipeline. Table 3 demonstrates how the amount of candidate words affects the accuracy of the pipeline. The second column, when the number of candidates is 1, corresponds to the ordinary dictionary-guided method. As can be observed, a substantial improvement of 2% in accuracy occurs when the candidate number increases from 1 to 10, which explains that the correct word is not necessarily the one with the smallest edit distance. The average accuracy marginally improves as the number of candidates increases from 10 to 80. The saturation appears when the number arrives at 150. Table 3 illustrates the primary benefit of the proposed method, which can select the correct output from a group of options.

Table 4. Comparison of recognition accuracy on different numbers of resemblant words.

Number of resemblant words	0	3	7	15	31
Recognition Accuracy	91.1	93.8	93.9	93.9	93.9

The Discussion of Resemblant Word Function: For image-text pairs to be successfully matched, a certain number of hard negatives are included in the training process. To illustrate the efficacy of this strategy, we arrange a variety of resemblant words: 0, 3, 7, 15 and 31. The recognition accuracy of the entire pipeline is shown in Table 4. The second column 0 indicates that no hard negative is used. It can be seen that recognition accuracy improves as the number of hard negatives increases. However, it will not be improved until a certain number of hard negatives has been accumulated. In contrast, when the number of hard negatives is equal to 0, the SITM network cannot complete the image-text matching task, therefore some incorrect matching pairs are produced. The performance(91.1% accuracy) is significantly worse than the ordinary dictionary method performance(92.1% accuracy). Table 4 demonstrates that the model is capable of learning more fine-grained distinctions between different text features through resemblant word generation strategy.

5   Conclusion

In this paper, we propose a new dictionary-guided scene text recognition method, which integrates the visual features into the inference stage and can effectively boost the performance of dictionary language model. In addition, the SITM is designed to indicate the correctness of explicit language model rectification. The resemblant words generation strategy, which utilizes labels to generate hard negatives in the training stage, is presented to improve the matching accuracy of SITM network. The experiments on six mainstream benchmarks demonstrate that our method outperforms the ordinary dictionary method and also show superiority in other state-of-the-art scene text recognition methods.

Acknowledgement. This work is supported by the National Natural Science Foundation of China under Grant No. 62176091.

References

1. Aberdam, A., et al.: Sequence-to-sequence contrastive learning for text recognition. In: Proceedings of the IEEE/CVF Conference on Computer Vision and Pattern Recognition, pp. 15302–15312 (2021)
2. Antol, S., et al.: VQA: visual question answering. In: Proceedings of the IEEE International Conference on Computer Vision, pp. 2425–2433 (2015)
3. Baek, J., et al.: What is wrong with scene text recognition model comparisons? Dataset and model analysis. In: Proceedings of the IEEE/CVF International Conference on Computer Vision, pp. 4715–4723 (2019)
4. Bautista, D., Atienza, R.: Scene text recognition with permuted autoregressive sequence models. In: idan, S., Brostow, G., Cissé, M., Farinella, G.M., Hassner, T. (eds.) Computer Vision. ECCV 2022. LNCS, vol. 13688, pp. 178–196. Springer, Cham (2022). https://doi.org/10.1007/978-3-031-19815-1_11
5. Cheng, Z., Bai, F., Xu, Y., Zheng, G., Pu, S., Zhou, S.: Focusing attention: towards accurate text recognition in natural images. In: Proceedings of the IEEE International Conference on Computer Vision, pp. 5076–5084 (2017)

6. Cheng, Z., Xu, Y., Bai, F., Niu, Y., Pu, S., Zhou, S.: Aon: towards arbitrarily-oriented text recognition. In: Proceedings of the IEEE Conference on Computer Vision and Pattern Recognition, pp. 5571–5579 (2018)
7. Fang, S., Xie, H., Wang, Y., Mao, Z., Zhang, Y.: Read like humans: autonomous, bidirectional and iterative language modeling for scene text recognition. In: Proceedings of the IEEE/CVF Conference on Computer Vision and Pattern Recognition, pp. 7098–7107 (2021)
8. Graves, A., Fernández, S., Gomez, F., Schmidhuber, J.: Connectionist temporal classification: labelling unsegmented sequence data with recurrent neural networks. In: Proceedings of the 23rd International Conference on Machine Learning, pp. 369–376 (2006)
9. Gupta, A., Vedaldi, A., Zisserman, A.: Synthetic data for text localisation in natural images. In: Proceedings of the IEEE Conference on Computer Vision and Pattern Recognition, pp. 2315–2324 (2016)
10. He, P., Huang, W., Qiao, Y., Loy, C.C., Tang, X.: Reading scene text in deep convolutional sequences. In: Thirtieth AAAI conference on artificial intelligence (2016)
11. Hu, W., Cai, X., Hou, J., Yi, S., Lin, Z.: GTC: Guided training of CTC towards efficient and accurate scene text recognition. In: Proceedings of the AAAI Conference on Artificial Intelligence, vol. 34, pp. 11005–11012 (2020)
12. Jaderberg, M., Simonyan, K., Vedaldi, A., Zisserman, A.: Synthetic data and artificial neural networks for natural scene text recognition. arXiv preprint arXiv:1406.2227 (2014)
13. Jaderberg, M., Simonyan, K., Vedaldi, A., Zisserman, A.: Deep structured output learning for unconstrained text recognition. In: ICLR (2015)
14. Jaderberg, M., Vedaldi, A., Zisserman, A.: Deep features for text spotting. In: Fleet, D., Pajdla, T., Schiele, B., Tuytelaars, T. (eds.) ECCV 2014. LNCS, vol. 8692, pp. 512–528. Springer, Cham (2014). https://doi.org/10.1007/978-3-319-10593-2_34
15. Jia, C., et al.: Scaling up visual and vision-language representation learning with noisy text supervision. In: International Conference on Machine Learning, pp. 4904–4916. PMLR (2021)
16. Karatzas, Det al.: ICDAR 2015 competition on robust reading. In: 2015 13th International Conference on Document Analysis and Recognition (ICDAR), pp. 1156–1160. IEEE (2015)
17. Karatzas, D., et al.: ICDAR 2013 robust reading competition. In: 2013 12th International Conference on Document Analysis and Recognition, pp. 1484–1493. IEEE (2013)
18. Lee, C.Y., Osindero, S.: Recursive recurrent nets with attention modeling for OCR in the wild. In: Proceedings of the IEEE Conference on Computer Vision and Pattern Recognition, pp. 2231–2239 (2016)
19. Levenshtein, V.I., et al.: Binary codes capable of correcting deletions, insertions, and reversals. In: Soviet Physics Doklady, vol. 10, pp. 707–710. Soviet Union (1966)
20. Li, H., Wang, P., Shen, C., Zhang, G.: Show, attend and read: a simple and strong baseline for irregular text recognition. In: Proceedings of the AAAI Conference on Artificial Intelligence, vol. 33, pp. 8610–8617 (2019)
21. Li, L., Yatskar, M., Yin, D., Hsieh, C., Chang, K.: A simple and performant baseline for vision and language. arXiv preprint arXiv:1908.03557 (2019)
22. Li, Y., Qi, H., Dai, J., Ji, X., Wei, Y.: Fully convolutional instance-aware semantic segmentation. In: Proceedings of the IEEE Conference on Computer Vision and Pattern Recognition, pp. 2359–2367 (2017)

23. Liao, M., et al.: Scene text recognition from two-dimensional perspective. In: Proceedings of the AAAI Conference on Artificial Intelligence, vol. 33, pp. 8714–8721 (2019)
24. Lu, J., Batra, D., Parikh, D., Lee, S.: ViLBERT: pretraining task-agnostic visiolinguistic representations for vision-and-language tasks. In: Advances in Neural Information Processing Systems, vol. 32 (2019)
25. Lu, J., Goswami, V., Rohrbach, M., Parikh, D., Lee, S.: 12-in-1: Multi-task vision and language representation learning. In: Proceedings of the IEEE/CVF Conference on Computer Vision and Pattern Recognition, pp. 10437–10446 (2020)
26. Luo, C., Jin, L., Chen, J.: SIMAN: exploring self-supervised representation learning of scene text via similarity-aware normalization. In: Proceedings of the IEEE/CVF Conference on Computer Vision and Pattern Recognition,pp. 1039–1048 (2022)
27. Lyu, P., Liao, M., Yao, C., Wu, W., Bai, X.: Mask TextSpotter: an end-to-end trainable neural network for spotting text with arbitrary shapes. In: Ferrari, V., Hebert, M., Sminchisescu, C., Weiss, Y. (eds.) Computer Vision – ECCV 2018. LNCS, vol. 11218, pp. 71–88. Springer, Cham (2018). https://doi.org/10.1007/978-3-030-01264-9_5
28. Merity, S., Xiong, C., Bradbury, J., Socher, R.: Pointer sentinel mixture models. arXiv preprint arXiv:1609.07843 (2016)
29. Mishra, A., Alahari, K., Jawahar, C.: Top-down and bottom-up cues for scene text recognition. In: 2012 IEEE Conference On Computer Vision and Pattern Recognition, pp. 2687–2694. IEEE (2012)
30. Mou, Y., Tan, L., Yang, H., Chen, J., Liu, L., Yan, R., Huang, Y.: PlugNet: degradation aware scene text recognition supervised by a pluggable super-resolution unit. In: Vedaldi, A., Bischof, H., Brox, T., Frahm, J.-M. (eds.) ECCV 2020. LNCS, vol. 12360, pp. 158–174. Springer, Cham (2020). https://doi.org/10.1007/978-3-030-58555-6_10
31. Nguyen, N., et al.: Dictionary-guided scene text recognition. In: Proceedings of the IEEE/CVF Conference on Computer Vision and Pattern Recognition, pp. 7383–7392 (2021)
32. Phan, T.Q., Shivakumara, P., Tian, S., Tan, C.L.: Recognizing text with perspective distortion in natural scenes. In: Proceedings of the IEEE International Conference on Computer Vision, pp. 569–576 (2013)
33. Radford, A., et al.: Learning transferable visual models from natural language supervision. In: International Conference on Machine Learning, pp. 8748–8763. PMLR (2021)
34. Risnumawan, A., Shivakumara, P., Chan, C.S., Tan, C.L.: A robust arbitrary text detection system for natural scene images. Expert Syst. App. 41(18), 8027–8048 (2014)
35. Sheng, F., Chen, Z., Xu, B.: NRTR: a no-recurrence sequence-to-sequence model for scene text recognition. In: 2019 International conference on document analysis and recognition (ICDAR), pp. 781–786. IEEE (2019)
36. Shi, B., Wang, X., Lyu, P., Yao, C., Bai, X.: Robust scene text recognition with automatic rectification. In: Proceedings of the IEEE Conference on Computer Vision and Pattern Recognition, pp. 4168–4176 (2016)
37. Shi, B., Yang, M., Wang, X., Lyu, P., Yao, C., Bai, X.: Aster: an attentional scene text recognizer with flexible rectification. IEEE Trans. Pattern Anal. Mach. Intell. 41(9), 2035–2048 (2018)
38. Su, B., Lu, S.: Accurate recognition of words in scenes without character segmentation using recurrent neural network. Pattern Recogn. 63, 397–405 (2017)

39. Su, W., et al.: Pre-training of generic visual-linguistic representations. In: Proceedings of the 8th International Conference on Learning Representations, pp. 1–14 (2020)

40. Suhr, A., Zhou, S., Zhang, A., Zhang, I., Bai, H., Artzi, Y.: A corpus for reasoning about natural language grounded in photographs. arXiv preprint arXiv:1811.00491 (2018)

41. Wan, Z., He, M., Chen, H., Bai, X., Yao, C.: TextScanner: reading characters in order for robust scene text recognition. In: Proceedings of the AAAI Conference on Artificial Intelligence, vol. 34, pp. 12120–12127 (2020)

42. Wang, K., Babenko, B., Belongie, S.: End-to-end scene text recognition. In: 2011 International Conference on Computer Vision, pp. 1457–1464. IEEE (2011)

43. Wang, P., Yang, L., Li, H., Deng, Y., Shen, C., Zhang, Y.: A simple and robust convolutional-attention network for irregular text recognition. arXiv preprint arXiv:1904.01375 6(2), 1 (2019)

44. Wang, Y., Xie, H., Fang, S., Wang, J., Zhu, S., Zhang, Y.: From two to one: a new scene text recognizer with visual language modeling network. In: Proceedings of the IEEE/CVF International Conference on Computer Vision, pp. 14194–14203 (2021)

45. Wojna, Z., et al.: Attention-based extraction of structured information from street view imagery. In: 2017 14th IAPR International Conference on Document Analysis and Recognition (ICDAR), vol. 1, pp. 844–850. IEEE (2017)

46. Xing, L., Tian, Z., Huang, W., Scott, M.R.: Convolutional character networks. In: Proceedings of the IEEE/CVF International Conference on Computer Vision, pp. 9126–9136 (2019)

47. Yang, M., et al.: Reading and writing: discriminative and generative modeling for self-supervised text recognition. In: Proceedings of the 30th ACM International Conference on Multimedia, pp. 4214–4223 (2022)

48. Yang, X., He, D., Zhou, Z., Kifer, D., Giles, C.L.: Learning to read irregular text with attention mechanisms. In: IJCAI, vol. 1, p. 3 (2017)

49. Yu, D., et al.: Towards accurate scene text recognition with semantic reasoning networks. In: Proceedings of the IEEE/CVF Conference on Computer Vision and Pattern Recognition, pp. 12113–12122 (2020)

50. Yue, X., Kuang, Z., Lin, C., Sun, H., Zhang, W.: RobustScanner: dynamically enhancing positional clues for robust text recognition. In: Vedaldi, A., Bischof, H., Brox, T., Frahm, J.-M. (eds.) ECCV 2020. LNCS, vol. 12364, pp. 135–151. Springer, Cham (2020). https://doi.org/10.1007/978-3-030-58529-7_9

51. Zhang, H., Yao, Q., Yang, M., Xu, Y., Bai, X.: AutoSTR: efficient backbone search for scene text recognition. In: Vedaldi, A., Bischof, H., Brox, T., Frahm, J.-M. (eds.) ECCV 2020. LNCS, vol. 12369, pp. 751–767. Springer, Cham (2020). https://doi.org/10.1007/978-3-030-58586-0_44

E2TIMT: Efficient and Effective Modal Adapter for Text Image Machine Translation

Cong Ma[1,2], Yaping Zhang[1,2(✉)], Mei Tu[4], Yang Zhao[1,2], Yu Zhou[2,3], and Chengqing Zong[1,2]

[1] School of Artificial Intelligence, University of Chinese Academy of Sciences, Beijing 100049, People's Republic of China
{cong.ma,yaping.zhang,yang.zhao,yzhou,cqzong}@nlpr.ia.ac.cn
[2] State Key Laboratory of Multimodal Artificial Intelligence Systems (MAIS), Institute of Automation, Chinese Academy of Sciences, Beijing 100190, People's Republic of China
[3] Fanyu AI Laboratory, Zhongke Fanyu Technology Co., Ltd, Beijing 100190, People's Republic of China
[4] Samsung Research China - Beijing (SRC-B), Beijing, China
mei.tu@samsung.com

Abstract. Text image machine translation (TIMT) aims to translate texts embedded in images from one source language to another target language. Existing methods, both two-stage cascade and one-stage end-to-end architectures, suffer from different issues. The cascade models can benefit from the large-scale optical character recognition (OCR) and MT datasets but the two-stage architecture is redundant. The end-to-end models are efficient but suffer from training data deficiency. To this end, in our paper, we propose an end-to-end TIMT model fully making use of the knowledge from existing OCR and MT datasets to pursue both an effective and efficient framework. More specifically, we build a novel modal adapter effectively bridging the OCR encoder and MT decoder. End-to-end TIMT loss and cross-modal contrastive loss are utilized jointly to align the feature distribution of the OCR and MT tasks. Extensive experiments show that the proposed method outperforms the existing two-stage cascade models and one-stage end-to-end models with a lighter and faster architecture. Furthermore, the ablation studies verify the generalization of our method, where the proposed modal adapter is effective to bridge various OCR and MT models. (Our codes are available at: https://github.com/EriCongMa/E2TIMT)

Keywords: Text image machine translation · Modal adapter · Cross modal contrastive learning

1 Introduction

Text image machine translation (TIMT) is the core research of many applications, such as scene text translation, document image translation, and photo

G. A. Fink et al. (Eds.): ICDAR 2023, LNCS 14192, pp. 70–88, 2023.
https://doi.org/10.1007/978-3-031-41731-3_5

translation. Approaches to TIMT are mainly divided into two categories: two-stage cascade method [1, 3, 7, 10, 26] and one-stage end-to-end method [5, 20, 29]. The cascade model deploys recognition and translation models sequentially, which benefits from training with existing large-scale optical character recognition (OCR) and machine translation (MT) datasets. However, the task gap between OCR and MT models might hurt the performance because translation models are vulnerable to recognition errors. Furthermore, the cascade model is two-stage, *i.e.* the sequential integration of OCR and MT models, thus is redundant in parameters and has a slow decoding speed. To alleviate the error propagation problem, some studies turn to exploring one-stage end-to-end architecture with fewer parameters and faster decoding speed [20]. However, the scarcity of end-to-end TIMT data limits the performance of end-to-end models. Although the multi-task learning enhanced end-to-end TIMT model incorporates external OCR datasets [5, 29] or MT datasets [19], the huge potential of fully benefiting from the knowledge of existing OCR and MT datasets or their corresponding pre-trained models is seldom explored. RTNet [29] is proposed to link the OCR encoder and MT decoder, but it ignores the task gap between recognition and translation tasks, causing limited performance. In summary, the following three major challenges are usually faced in the TIMT study:

- **Task Gap.** There is a large domain gap between the OCR/MT tasks, which indicates the direct connection of the recognition and translation models is not optimal.
- **Cascade Redundancy.** It leads to model/complexity redundancy when directly cascading existing OCR and MT models without any optimization.
- **End-to-end Data Scarcity.** The dataset for end-to-end TIMT is scarce. It is critical to transfer knowledge from existing OCR and MT datasets or pre-trained models, which is seldom explored by previous methods.

In this paper, we propose a novel modal adapter architecture to improve the end-to-end TIMT model by eliminating task gaps and making full of the knowledge from pre-trained OCR and MT models. Furthermore, the modal adapter can is a parameter efficient fine-tuning method, which just optimizes parameters of modal adapter by frozen pre-trained encoder and decoder. Thus, modal adapter based TIMT model has much fewer parameters to update compared with end-to-end models and has a faster inference speed than cascade models. In detail, a self-attention based modal adapter is incorporated between the pre-trained OCR encoder and MT decoder. Different from vanilla adapter tuning [24], which is just fine-tuned on downstream tasks, the task gap is bridged in our framework by a cross-modal contrastive loss that aligns the distributions between the OCR and MT features of the same sentence content. Two types of modal adapters are studied to validate the effectiveness of bridging various OCR and MT modules. Embedding modal adapter (EmbMA) is proposed to bridge OCR image encoder and MT sequential encoder, while sequential modal adapter (SeqMA) is inserted between OCR Sequential encoder and MT decoder. Finally, the MT decoder generates the translation from the features transformed by the modal adapter. Our contributions are summarized as follows:

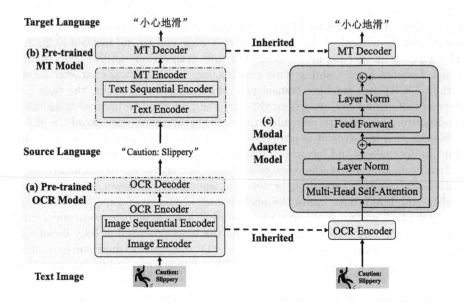

Fig. 1. Architectures of OCR, MT, and Modal Adapter for TIMT Model. The solid arrow lines represent the data flow in the model. The dotted arrow lines denote the parameters of encoder and decoder in modal adapter based TIMT are inherited from pre-trained OCR/MT models.

- We propose a modal adapter based TIMT model to unify cascade and end-to-end models by bridging the pre-trained recognition encoder and translation decoder.
- Cross-modal contrastive learning is incorporated to align the distribution of image features and text features encoded by an OCR encoder and an MT encoder respectively, which alleviates the OCR-MT task gap and improves the performance of text image machine translation.
- Extensive experiments show our method outperforms both the existing cascade models and end-to-end models with a lighter and faster architecture. Furthermore, the modal adapter has a good generalization when bridging various recognition encoders and translation decoders.

2 Preliminary

To unify the processing progress of recognition and translation models, we divide both the OCR and MT encoders into two submodules: image/text encoder for embedding encoding, and sequential encoder for contextual feature extraction. We will introduce the processing flow of OCR and MT models individually.

2.1   OCR Model

As shown in Fig. 1 (a), given an input image I, a convolutional neural network (CNN) based image encoder extracts the image embedding E_I by transforming image pixels into feature vectors:

$$E_I = \text{CNN}(I) \tag{1}$$

where $I \in \mathbb{R}^{H \times W \times C}$ and $E_I \in \mathbb{R}^{l_I \times c}$. H, W, and C denote the height, width, and channel of the input image respectively. l_I represents the length of image embedding, which is calculated as $l_I = h \times w$, where h, w, and c denote the height, width, and channel of the encoded feature map separately.

The image encoder mainly extracts the local features of the input images, while the image sequential encoder aims to model contextual information by considering the whole input sequence:

$$S_I = \text{Seq}_I(E_I) \tag{2}$$

where $\text{Seq}_I(\cdot)$ represents the image sequential encoder and transformer encoder [32] is utilized in our implementation. $S_I \in \mathbb{R}^{l_S \times d_S}$ denotes the sequential features in OCR model. l_S and d_S denote the length and dimension of sequential features respectively.

Finally, the OCR decoder generates recognized tokens auto-regressively given sequential features:

$$\begin{aligned} D_I &= \text{Dec}_I(S_I); \\ P(X|I) &= \text{Softmax}(W_I D_I) \end{aligned} \tag{3}$$

where $\text{Dec}_I(\cdot)$ represents the OCR decoder, and transformer decoder [32] is utilized in our implementation. D_I denotes the outputs of the decoder. $W_I \in \mathbb{R}^{|\mathcal{V}_X| \times d_I}$ represents the linear transformation that maps the decoder features into corresponding recognized tokens, $\mathcal{V}_X$ is the recognition vocabulary, and d_I is the dimension of decoder hidden states.

2.2   MT Model

MT model translates the source language into the target language as shown in Fig. 1 (b). Given a source language sentence T, the text encoder first maps the input words into a sequence of word embeddings:

$$E_T = \text{Embedding}(T) \tag{4}$$

where $E_T \in \mathbb{R}^{l_E \times d_E}$ denotes the text embedding. l_E and d_E represent the sequence length and the dimension of text embedding respectively.

Text sequential encoder further extracts contextual features based on text embeddings:

$$S_T = \text{Seq}_T(E_T) \tag{5}$$

where $\text{Seq}_T(\cdot)$ represents the text sequential encoder, which is a transformer encoder in our implementation. S_T denotes the encoded text sequential features.

MT decoder finally generates the target tokens auto-regressively given sequential features:

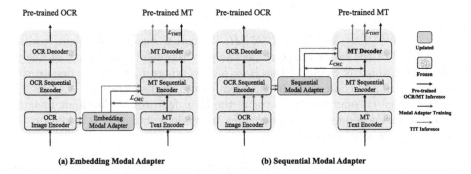

(a) Embedding Modal Adapter   **(b) Sequential Modal Adapter**

Fig. 2. Diagram of (a) Embedding Modal Adapter and (b) Sequential Modal Adapter. Black, red and blue arrow lines denote the pre-trained OCR/MT, modal adapter training and TIMT inference flows respectively. The green box refers to trainable parameters and the blue box refers to frozen ones. (Color figure online)

$$D_T = \text{Dec}_T(S_T);$$
$$P(Y|T) = \text{Softmax}(W_T D_T) \tag{6}$$

where $\text{Dec}_T(\cdot)$ represents the MT decoder, and the transformer decoder is utilized in our implementation. D_T is the output of the decoder and $W_T \in \mathbb{R}^{|\mathcal{V}_Y| \times d_T}$ is the linear transformation. $\mathcal{V}_Y$ represents the target language vocabulary and d_T denotes the dimension of the hidden states.

3   Methodology

To bridge the pre-trained OCR encoder and the MT decoder, the modal adapter is proposed to transform the OCR features into the MT feature space as shown in Fig. 1 (c).

Specifically, features of the OCR encoder are transformed by the stacked modal adapter layer:

$$\hat{H}_{\text{MA}}^n = \text{LN}(\text{MSA}(H_{\text{MA}}^{n-1})) + H_{\text{MA}}^{n-1}$$
$$H_{\text{MA}}^n = \text{LN}(\text{FFN}(\hat{H}_{\text{MA}}^n)) + \hat{H}_{\text{MA}}^n \tag{7}$$

where H_{MA}^n denotes the output of the n-th modal adapter layer, and H_{MA}^0 is the feature from the OCR encoder. $\text{MSA}(\cdot)$, $\text{FFN}(\cdot)$, and $\text{LN}(\cdot)$ represent multihead self-attention, feed-forward, and layer norm modules respectively. After transformation by the modal adapter, features encoded by the OCR encoder are further fed into the MT decoder to generate translation results.

Since there are two submodules in the OCR encoder (image encoder and sequential encoder) as introduced in Sect. 2.1, we propose two types of modal adapters. The first one is the embedding modal adapter (EmbMA), which aims at aligning the image embedding and text embedding. The second one is the sequential modal adapter (SeqMA), which transforms the sequential features encoded by the image sequential encoder to the sequential feature space of the MT task. We will introduce our proposed EmbMA and SeqMA in detail.

3.1 Embedding Modal Adapter

The embedding modal adapter is placed in the middle of the OCR image encoder and the text sequential encoder as shown in Fig. 2 (a). First, the EmbMA transforms the image embedding into the text embedding space. Second, to better meet the feature distribution of the MT processing flow, the output of EmbMA is constrained by the text embedding through a cross-modal contrastive loss $\mathcal{L}_{\mathrm{CMC}}^{\mathrm{EmbMA}}$. As so, the output of EmbMA given i-th image embedding should be similar to the i-th text embedding, and apart from the other text embeddings in the mini-batch:

$$H_{\mathrm{EmbMA}}^{(i)} = \mathrm{EmbMA}(E_I^{(i)})$$

$$\mathcal{L}_{\mathrm{CMC}}^{\mathrm{EmbMA}} = -\sum_{i=1}^{K} \log \frac{\exp(d(H_{\mathrm{EmbMA}}^{(i)}, E_T^{(i)})/\tau)}{\sum_{j=1}^{K} \exp(d(H_{\mathrm{EmbMA}}^{(i)}, E_T^{(j)})/\tau)} \quad (8)$$

where $\mathrm{EmbMA}(\cdot)$ utilizes the same modal adapter architecture as in Eq. 7. $H_{\mathrm{EmbMA}}^{(i)}$ represents the output of the EmbMA. $E_I^{(i)}$ and $E_T^{(i)}$ denote the image and text embedding of i-th sample respectively. K denotes the size of the mini-batch. τ stands for the temperature parameter and $d(q, k)$ represents the similarity metric which we utilize cosine similarity in our implementation.

Aligned with text embedding, the outputs of EmbMA are further fed into the text sequential encoder to obtain the contextual feature $S_{\mathrm{EmbMA}}^{(i)}$. Through EmbMA, the image embeddings are transformed into the MT processing flow, and MT decoder finally generates target translation:

$$S_{\mathrm{EmbMA}}^{(i)} = \mathrm{Seq}_T(H_{\mathrm{EmbMA}}^{(i)})$$

$$D_{\mathrm{EmbMA}}^{(i)} = \mathrm{Dec}_T(S_{\mathrm{EmbMA}}^{(i)}) \quad (9)$$

$$P(Y^{(i)}|I^{(i)}) = \mathrm{Softmax}(W_T D_{\mathrm{EmbMA}}^{(i)})$$

3.2 Sequential Modal Adapter

Different from EmbMA, SeqMA is designed to align the sequential features of the OCR and MT models. As shown in Fig. 2 (b), SeqMA first transforms the image sequential features into text sequential feature space. Then, the MT decoder generates target language tokens given transformed image sequential features:

$$H_{\text{SeqMA}}^{(i)} = \text{SeqMA}(S_I^{(i)})$$
$$D_{\text{SeqMA}}^{(i)} = \text{Dec}_T(H_{\text{SeqMA}}^{(i)}) \tag{10}$$
$$P(Y^{(i)}|I^{(i)}) = \text{Softmax}(W_T D_{\text{SeqMA}}^{(i)})$$

where SeqMA$(\cdot)$ uses the same structure as in Eq. 7. $H_{\text{SeqMA}}^{(i)}$ denotes the output of the sequential modal adapter and $S_I^{(i)}$ represents the output of the image sequential encoder of the i-th sample in the mini-batch. $D_{\text{SeqMA}}^{(i)}$ denotes the output of text decoder given the hidden states from SeqMA.

Since the hidden states of the SeqMA are further fed into the MT decoder, the feature distribution of $H_{\text{SeqMA}}^{(i)}$ should be similar to the hidden states of $S_T^{(i)}$. To bridge the feature gap between the OCR and MT tasks, a cross-modal contrastive loss is utilized to align the feature distribution of the transformed image sequential feature and text sequential feature:

$$\mathcal{L}_{\text{CMC}}^{\text{SeqMA}} = -\sum_{i=1}^{K} \log \frac{\exp(d(H_{\text{SeqMA}}^{(i)}, S_T^{(i)})/\tau)}{\sum_{j=1}^{K} \exp(d(H_{\text{SeqMA}}^{(i)}, S_T^{(j)})/\tau)} \tag{11}$$

where $d(\cdot)$ and τ are the same similarity metric and temperature parameter as introduced in Eq. 8.

3.3   Training of Modal Adapter

During model training, only parameters in modal adapters are updated, while the parameters in the OCR and MT models are all fixed. Through parameter-efficient modal adapter tuning, the pre-trained OCR encoder and MT decoder are able to transfer to the TIMT task with ease. Specifically, multi-task learning is utilized by optimizing end-to-end text image translation loss and cross-modal contrastive loss. Formally, the end-to-end text image translation loss and the overall loss functions are:

$$\mathcal{L}_{\text{TIMT}} = -\sum_{i=1}^{|D_{\text{TIMT}}|} \log P(Y^{(i)}|I^{(i)}) \tag{12}$$

$$\mathcal{L}_{\text{All}} = (1 - \lambda_{\text{CMC}})\mathcal{L}_{\text{TIMT}} + \lambda_{\text{CMC}}\mathcal{L}_{\text{CMC}}$$

where $\mathcal{L}_{\text{CMC}}$ is introduced as in Eq. 8 and Eq. 11. λ_{CMC} denotes the hyper-parameter, which balances the weight of end-to-end text image translation loss and cross-modal contrastive loss. Note that the $P(Y^{(i)}|I^{(i)})$ in end-to-end text image translation loss $\mathcal{L}_{\text{TIMT}}$ and cross-modal contrastive loss $\mathcal{L}_{\text{CMC}}$ are calculated based on the corresponding training workflow of SeqMA and EmbMA.

3.4   Inference

During model inference, as the blue arrow lines shown in Fig. 2, the input images are first fed into the OCR encoder to obtain the image features. Second, the modal adapter transforms the image features into the MT feature space, and the MT decoder finally generates translation results. Note that the OCR decoder and the MT encoder are not utilized during inference resulting in a fast decoding speed with the end-to-end processing architecture as shown in Fig. 1 (c). By bridging OCR encoder and MT decoder, modal adapter based method can take full advantage of pre-trained OCR and MT models.

4   Experiments

4.1   Datasets

OCR, MT, and end-to-end TIMT datasets are utilized in our experiments. OCR and MT datasets are used to train the OCR and MT models respectively. While the TIMT dataset is used to train the parameters in the modal adapter.

OCR Datasets. OCR datasets are composed of text images and corresponding text pairs $\{(I_i, T_i)\}_{i=1}^{|D_{\mathrm{OCR}}|}$. Three OCR datasets are considered in our experiments. **MJSynth (MJ)** [12][1] is a synthetic word box image recognition dataset designed for English scene text recognition containing 8.9M synthetic word box images. **SynthText (ST)** [9][2] is another synthetic dataset containing 5.5M word box images, which renders the texts onto real-world scene images. **Synthetic Text Line Dataset** is a customized text line recognition dataset that is constructed with the rule-based synthetic method[3]. 1M English and 1M Chinese synthetic text line recognition pairs are synthesized in our experiments.

MT Datasets. Parallel sentences from the Workshop of Machine Translation 2018[4] are utilized to train the text machine translation models. Specifically, three translation directions are considered in our experiments: English-to-Chinese (En⇒Zh), English-to-German (En⇒De), and Chinese-to-English (Zh⇒En). After pre-processing and filtering, 5,984,287 En⇔Zh and 20,895,771 En⇒De translation pairs are finally obtained to train MT models.

End-to-End TIMT Datasets. A public end-to-end TIMT dataset proposed by [19] is utilized to train end-to-end TIMT models. This dataset is a synthetic text image translation corpus by synthesizing the text image through a rule-based toolkit given randomly selected background images, font types, and other rendering effects, which is similar to the synthesis method as synthetic text line recognition dataset. The parallel sentences of the end-to-end text image translation datasets are extracted from the text translation corpus. In summary, one million end-to-end TIMT pairs are utilized for each translation direction.

---

[1] https://www.robots.ox.ac.uk/vgg/data/text/.
[2] https://www.robots.ox.ac.uk/vgg/data/scenetext/.
[3] https://github.com/Belval/TextRecognitionDataGenerator.
[4] http://www.statmt.org/wmt18/.

Evaluation Datasets. Evaluation sets constructed by [19] are used to measure the performance of various models. Three domains are considered, including synthetic, subtitle, and street-view evaluation domains. The synthetic evaluation dataset contains 2,502 En⇔Zh and 2,000 En⇒De translation pairs, which are synthesized as the synthetic training dataset. For real-world evaluation datasets, the En⇔Zh subtitle dataset contains 1,040 translation pairs, while the En⇒Zh street-view dataset contains 1,198 translation pairs.

Table 1. Comparison of end-to-end, cascade and modal adapter tuning based text image machine translation models.

Architecture	Synthetic			Subtitle		Street
	En⇒Zh	En⇒De	Zh⇒En	En⇒Zh	Zh⇒En	Zh⇒En
End-to-End Models						
TRBA [2]	9.61	7.36	4.77	12.12	5.18	0.36
CLTIR [5]	18.02	15.55	10.74	16.47	9.04	0.43
CLTIR+OCR [5]	19.44	16.31	13.52	17.96	11.25	1.74
RTNet [29]	18.91	15.82	12.54	17.63	10.63	1.07
RTNet+OCR [29]	19.63	16.78	14.01	18.82	11.50	1.93
MTETIMT [19]	19.25	16.27	13.16	17.73	10.79	1.69
MTETIMT+MT [19]	21.96	18.84	15.62	19.17	12.11	5.84
MHCMM [4]	22.08	18.97	15.66	19.24	12.12	5.87
Cascade Models						
CRNN + Transformer	14.43	11.27	10.52	17.88	10.06	3.25
TRBA + Transformer	17.59	13.86	12.79	18.22	10.53	4.08
TRT + Transformer	20.46	16.48	15.12	19.12	12.08	5.78
Modal Adapter Tuning Models						
Sequential Modal Adapter	20.90	19.02	15.22	19.31	12.03	5.81
Embedding Modal Adapter	**22.53**	**19.67**	**16.25**	**19.46**	**12.39**	**6.24**

4.2 Experimental Settings

We implement the image encoder based on the code release by [2]. The MT model is utilized the same architecture proposed in [32]. The OCR and MT models are firstly trained with OCR and MT datasets respectively. Parameters of OCR and MT models are then frozen during fine-tuning. The implementation of the modal adapter is utilized a similar architecture as the transformer encoder with the hidden dimensions of 512, 8 attention heads, and a dropout rate of 0.1. The initial learning rate is set to 2e-3, the batch size is 64, and the training step is set to 300,000. Parameters of the modal adapter are initialized with Xavier initiation method [8] and optimized with Adam optimizer [15] on single NVIDIA V100 GPU. Detokenized BLEU [21] calculated by sacre-BLEU[5] is utilized as the metric to evaluate the performance of text image translation models.

---

[5] https://github.com/mjpost/sacrebleu.

4.3   Comparison of Various Text Image Translation Models

Table 1 shows the BLEU scores of text image translation models on various evaluation datasets. Three OCR models are utilized in the cascade models: CRNN [27], TPS+ResNet+BiLSTM+Attention (TRBA) [2], and TPS+ResNet+Transformer (TRT). While transformer-base [32] is utilized for MT model. The performance of the OCR and the MT models are shown in Sect. 4.4. Five architectures are compared in end-to-end TIMT setting. TRBA [2] represents the OCR architecture trained with end-to-end TIMT dataset. CLTIR [5] model trains end-to-end TIMT with auxiliary OCR task. RTNet [29] utilizes a feature transformer to link OCR encoder and decoder but ignores the task gap modeling. MTETIMT [19] represents the machine translation enhanced end-to-end TIMT model, which utilizes multi-task learning with auxiliary translation task. While MHCMM [4] proposes a multi-hierarchy cross-modal mimic framework for the end-to-end text image translation, which incorporates external text translation corpus and utilizes text MT model as teacher guidance for TIMT model. The modal adapter in Table 1 bridges the pre-trained TRT OCR encoder and transformer MT decoder. Experimental results show that our proposed sequential and embedding modal adapter outperforms two-stage cascade models on three translation domains with an average improvement of 1.01 BLEU scores. Meanwhile, modal adapter improves the TIMT performance on various language directions (En⇒Zh and En⇒De), revealing the method is robust to different language settings. For Zh⇒En translation direction, modal adapter based method achieves similar results as the previous machine translation enhanced multi-task training model, indicating modal adapter method can take full advantage of the pre-trained MT model without multi-task training.

Table 2. Performance of text image recognition models. Metric of scene text recognition (Rec.) is word accuracy and character error rate is utilized for text line recognition evaluation. Tr.E and Tr.D represent transformer encoder and decoder respectively.

Architecture	Image Encoder	Image Sequential Encoder	Decoder	Scene Text Rec.			Text Line Recognition		
				IIIT 3000	SVT 647	SP 645	Synthetic 2502	Subtitle 1040	Street 1198
CRNN [27]	VGG	BiLSTM	CTC	81.3	79.0	66.7	13.90	4.95	56.82
TRBA [2]	ResNet	BiLSTM	Attention	86.6	87.8	76.9	12.29	3.01	51.67
TRT	ResNet	Tr.E	Tr.D	87.9	87.2	78.6	10.89	2.33	49.83

Table 3. Performance of text translation models. BLEU score is utilized as the metric of text translation task.

Architecture	Synthetic			Subtitle		Street
	En⇒Zh	En⇒De	Zh⇒En	En⇒Zh	Zh⇒En	Zh⇒En
Transformer-Base [32]	25.38	20.97	17.56	19.64	13.78	15.17
Transformer-Big [32]	26.41	22.15	19.04	20.39	14.66	16.93

Furthermore, the embedding modal adapter performs better than the sequential modal adapter, and we attribute that EmbMA retains the cross-attention flow between the original text sequential encoder and decoder. This shows it is vital not only to eliminate the gap between the OCR and MT tasks but also to maintain the consistency of structures within each task.

4.4 Performance of OCR and MT Models

OCR and MT models in cascade models are firstly trained with corresponding OCR and MT datasets. Parameters in pre-trained OCR and MT models are then frozen during modal adapter training. Three OCR models CRNN, TRBA, and TRT are all trained with the same scene text recognition and synthetic text line recognition datasets introduced in Sect. 4.1. Table 2 shows the performance of various OCR models. Transformer based TRT model achieves the best recognition performance, indicating the strong sequential encoder is essential for optical character recognition. For MT models, the transformer-base and the transformer-big [32] are utilized to translate the source language into the target language. Table 3 shows the performance of text translation, and the transformer-big achieves better translation BLEU.

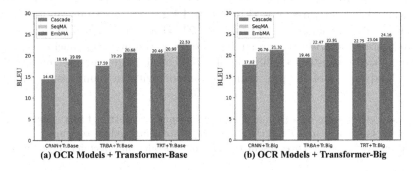

Fig. 3. Performance of various OCR and MT combinations with modal adapter. CRNN, TRBA, and TRT represent three OCR models. While MT Models include transformer-base (Tr.Base) and transformer-big (Tr.Big).

4.5 Generalization of Modal Adapter on Various OCR and MT Combinations

To evaluate the generalization of our proposed method, the modal adapter is studied by bridging various OCR encoders and MT decoders. As shown in Fig. 3, modal adapter tuning outperforms the cascade models on different OCR and MT combinations, revealing the good generalization of modal adapter tuning methods. Figure 3 (a) shows the text image translation results by combining different OCR models and transformer base MT model. Better OCR image encoder can extract more information into image features, leading to better text image translation performance.

Figure 3 (b) depicts various OCR models with transformer big MT models. Similar to Fig. 3 (a), better OCR models achieve better results with transformer big MT models. Furthermore, stronger MT decoders can further improve the translation performance in Fig. 3 (b) compared with Fig. 3 (a). As a result, our proposed modal adapter tuning method has strong scalability by bridging better OCR and MT models.

4.6 Analysis on Model Size and Decoding Speed of TIMT Models

Cascade models have redundant parameters and slow decoding speed. By removing the OCR decoder and the MT encoder, the modal adapter tuning method has fewer parameters and a faster decoding speed. As shown in Table 4, the end-to-end model, which is trained from the scratch, has fewer parameters and a faster decoding speed compared with the cascade model. Fine-tuning model is also an end-to-end model, which is initialized with the OCR encoder and MT decoder. Then the fine-tuning model is trained with the end-to-end text image translation dataset. Since the modal adapter bridges the OCR encoder and the MT decoder directly, it has a faster decoding speed than the cascade model. Meanwhile, after removing the OCR decoder and MT encoder, modal adapter models have fewer parameters than the cascade model. For the comparison of fine-tuning methods, modal adapter tuning outperforms fine-tuning model, because modal adapter models the task consistency between the OCR encoder and MT decoder, which alleviates the gap between OCR and MT tasks.

Table 4. Comparison of model size and decoding speed among various models on English-to-Chinese translation direction. The unit of parameters is million ($\times 10^6$), while the unit for speed is sentence per second. BLEU score is utilized to show the performance of synthetic and subtitle text image translation.

Architecture	Finetuned Params	Total Params	Speed	Synthetic	Subtitle
Cascade	-	195.1M	3.07	20.46	19.12
End-to-End	-	121.9M ($\downarrow$37.52%)	5.21 ($\uparrow$1.70x)	19.63	18.82
Fine-tuning	121.9M	121.9M ($\downarrow$37.52%)	5.21 ($\uparrow$1.70x)	20.18	19.04
SeqMA	13.2M	135.1M ($\downarrow$30.75%)	5.12 ($\uparrow$1.67x)	20.90	19.31
EmbMA				**22.53**	**19.46**

Table 5. Comparison of adapter tuning and modal adapter tuning on English-to-Chinese translation.

Architecture	Synthetic	Subtitle
Adapter Tuning	16.72	15.87
SeqMA (Bottleneck)	18.25	16.80
EmbMA (Bottleneck)	21.82	19.35
SeqMA	20.90	19.31
EmbMA	**22.53**	**19.46**

4.7   Comparison with Adapter Tuning

Adapter tuning [24] is an effective parameter-efficient fine-tuning method. Different from adapter tuning, which inserted bottleneck modules inside the pre-trained transformer layers, the modal adapter is designed outside the pre-trained models by bridging the separated OCR encoder and the MT decoder. As shown in Table 5, the modal adapter significantly outperforms adapter tuning with 5.81 BLEU for the synthetic domain and 3.59 BLEU for the subtitle domain. To offer a more similar architecture, we also put the bottleneck-based adapter outside the pre-trained models, which is similar to our proposed modal adapter tuning. Bottleneck-based modal adapter tuning also outperforms the vanilla adapter tuning, revealing the effectiveness of explicitly modeling the transformation mapping from the OCR feature space to the MT feature space. Finally, self-attention based modal adapter outperforms the bottleneck-based modal adapter, which we attribute to the strong encoding ability of stacked self-attention layers.

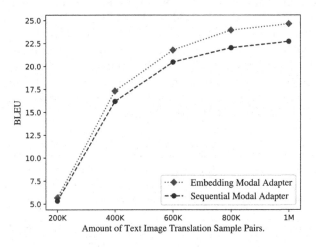

Fig. 4. Analysis on the amount of end-to-end TIT datasets on synthetic English-to-Chinese validation set.

4.8   Analysis on the Amount of End-to-End TIMT Dataset

Parameters of the modal adapter are trained on the end-to-end TIMT dataset and the amount of end-to-end data has a great impact on performance. Figure 4 shows the performance of modal adapter tuning with different amounts of end-to-end TIMT datasets. When the end-to-end data is low-resource (around 200 thousand image-text pairs), the performance of modal adapter tuning is limited. We attribute the reason to the non-convergence of modal adapter given low-resource end-to-end data. As the amount of end-to-end TIMT data increases, the modal adapter achieves better results, revealing the modal adapter needs enough data to learn the transformation from the OCR feature space to the

MT feature space. When the end-to-end image-text translation data achieves more than 800 thousand pairs, the TIMT results tend to be stable and perform the best translation results. Thus, one million end-to-end text image translation pairs are suitable to train a good end-to-end TIMT model.

4.9 Hyper-parameter Analysis

Hyper-parameter λ_{CMC} is an important parameter to balance the end-to-end TIMT optimization object and cross-modal contrastive learning object. Figure 5 shows the evaluation of hyper-parameter λ_{CMC}. From this hyper-parameter evaluation, the optimal value of λ_{CMC} is 0.4 for both embedding modal adapter and sequential modal adapter. When $\lambda_{CMC} = 0$, parameters in the modal adapter are only optimized by the end-to-end TIMT loss, which ignores the task gap between OCR and MT, leading to performance drop. Specifically, without cross-modal contrastive learning, SeqMA drops 2.35 BLEU scores and EmbMA drops 2.68 BLEU scores, indicating that cross-modal contrastive learning can effectively alleviate the feature gaps between the OCR and MT tasks. When $\lambda_{CMC} = 1$, the overall loss function becomes $\mathcal{L}_{All} = \mathcal{L}_{CMC}$, and the performance drops, indicating end-to-end loss is also vital to modal adapter tuning. Thus, the optimization of the modal adapter should be guided both from direct translation object $\mathcal{L}_{TIMT}$ and cross-modal contrastive learning object $\mathcal{L}_{CMC}$.

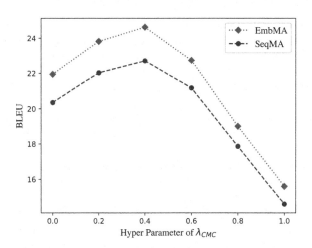

Fig. 5. Hyper-parameter evaluation of λ_{CMC} on English-to-Chinese validation set.

5   Related Work

5.1   Text Image Translation.

TIMT models are mainly divided into the cascade and end-to-end models. Cascade models deploy OCR and MT models respectively [1,3,7,10,26]. Specifically, the source language text images are first fed into OCR models to obtain the recognized source language sentences [2,13,14,27,28,35,36]. Second, the source language sentences are translated into the target language with the MT model [31,32,37,38]. Cascade directly connects separated OCR and MT models leading to model redundancy and slow decoding speed. Furthermore, recognition errors made by OCR models are further propagated through MT models, causing severe translation mistakes.

For end-to-end models, the naive approach is to take the OCR model to translate source language text images by training with source language images and corresponding target sentences like TRBA [2]. Furthermore, multi-task learning is proposed to incorporate external OCR datasets [5,29] or MT datasets [19] to enhance the performance of end-to-end models. MHCMM [4] further improves the feature representation through cross-modal mimic learning on the basis of incorporating external MT data.

However, existing methods still have limitations in fusing cascade and end-to-end models. In this paper, our proposed modal adapter bridges OCR encoder and MT decoder in cascade method through an end-to-end framework, which can take advantage of both cascade and end-to-end methods. Experimental results show modal adapter based TIMT effectively improves translation performance with efficient architecture and fast decoding speed.

5.2 Methods of Bridging Encoder and Decoder

Pre-trained models have been explored to achieve good performance after fine-tuning on down-stream tasks [6,17,22,23]. To simplify and speed up the fine-tuning process, efficiency tuning methods are proposed by just updating partial parameters of the model [34]. Another parameter-efficient tuning research keeps the parameters of pre-trained models unchanged and incorporates external modules to meet the downstream tasks like adapter tuning [24], LoRA [11], Bit-Fit [33], prefix tuning [18], and so on. These fine-tuning methods just optimize the parameters of external modules, which makes the fine-tuning process more efficient.

Except for fine-tuning unified pre-trained models, existing research also tried to bridge pre-trained encoder and decoder [25]. [30] proposed to bridge pre-trained mBERT and mGPT through a Graft module to achieve text machine translation. While [16] explores combining ASR encoder and MT decoder with vanilla adapter for end-to-end speech translation. Inspired by recent research on bridging encoder and decoder, we propose a modal adapter to bridge the OCR encoder and the MT decoder.

6   Conclusion

In this paper, we propose a faster and better modal adapter tuning method for the TIMT task, bridging the pre-trained OCR encoder and MT decoder. The sequential modal adapter and embedding adapter are evaluated to verify the effectiveness of bridging different OCR and MT modules. Extensive experiments show embedding modal adapter has better performance because it retains the cross-attention flow between the original MT sequential encoder and decoder. Meanwhile, with an end-to-end architecture, the modal adapter based method outperforms the cascade method with faster decoding speed and lightweight architecture. Furthermore, the modal adapter is effective to bridge various OCR and MT frameworks, revealing the good generalization of the modal adapter tuning method. In the next step, we will design more bridge modules for text image machine translation.

Acknowledgement. This work has been supported by the National Natural Science Foundation of China (NSFC) grants 62106265.

References

1. Afli, H., Way, A.: Integrating optical character recognition and machine translation of historical documents. In: Proceedings of the Workshop on Language Technology Resources and Tools for Digital Humanities, LT4DH@COLING, Osaka, Japan, December 2016, pp. 109–116 (2016)
2. Baek, J., et al.: What is wrong with scene text recognition model comparisons? Dataset and model analysis. In: 2019 IEEE/CVF International Conference on Computer Vision, ICCV 2019, Seoul, Korea (South), October 27 - November 2, 2019, pp. 4714–4722 (2019)
3. Chen, J., Cao, H., Natarajan, P.: Integrating natural language processing with image document analysis: what we learned from two real-world applications. Int. J. Doc. Anal. Recogn. (IJDAR) **18**(3), 235–247 (2015). https://doi.org/10.1007/s10032-015-0247-x
4. Chen, Z., Yin, F., Yang, Q., Liu, C.L.: Cross-lingual text image recognition via multi-hierarchy cross-modal mimic. IEEE Trans. Multimed. (TMM), 1–13 (2022)
5. Chen, Z., Yin, F., Zhang, X., Yang, Q., Liu, C.: Cross-lingual text image recognition via multi-task sequence to sequence learning. In: 25th International Conference on Pattern Recognition, ICPR 2020, Virtual Event/Milan, Italy, 10–15 January 2021, pp. 3122–3129 (2020)
6. Devlin, J., Chang, M., Lee, K., Toutanova, K.: BERT: pre-training of deep bidirectional transformers for language understanding. In: Proceedings of the 2019 Conference of the North American Chapter of the Association for Computational Linguistics: Human Language Technologies, NAACL-HLT 2019, Minneapolis, MN, USA, 2–7 June 2019, Volume 1 (Long and Short Papers), pp. 4171–4186. Association for Computational Linguistics (2019)
7. Du, J., Huo, Q., Sun, L., Sun, J.: Snap and translate using windows phone. In: 2011 International Conference on Document Analysis and Recognition, ICDAR 2011, Beijing, China, 18–21 September 2011, pp. 809–813. IEEE Computer Society (2011)
8. Glorot, X., Bengio, Y.: Understanding the difficulty of training deep feedforward neural networks. In: Proceedings of the Thirteenth International Conference on Artificial Intelligence and Statistics, AISTATS 2010, Chia Laguna Resort, Sardinia, Italy, May 13–15, 2010. JMLR Proceedings, vol. 9, pp. 249–256. JMLR.org (2010)
9. Gupta, A., Vedaldi, A., Zisserman, A.: Synthetic data for text localisation in natural images. In: 2016 IEEE Conference on Computer Vision and Pattern Recognition, CVPR 2016, Las Vegas, NV, USA, 27–30 June 2016, pp. 2315–2324. IEEE Computer Society (2016)

10. Hinami, R., Ishiwatari, S., Yasuda, K., Matsui, Y.: Towards fully automated manga translation. In: The Thirty-Fifth AAAI Conference on Artificial Intelligence, AAAI 2021, 2–9 February 2021 (2021)
11. Hu, E.J., et al.: Lora: low-rank adaptation of large language models. In: The Tenth International Conference on Learning Representations, ICLR 2022, Virtual Event, 25–29 April 2022. OpenReview.net (2022)
12. Jaderberg, M., Simonyan, K., Vedaldi, A., Zisserman, A.: Synthetic data and artificial neural networks for natural scene text recognition. CoRR abs/1406.2227 (2014)
13. Kaur, H., Kumar, M.: Offline handwritten Gurumukhi word recognition using extreme gradient boosting methodology. Soft. Comput. 25(6), 4451–4464 (2021)
14. Kaur, H., Kumar, M.: On the recognition of offline handwritten word using holistic approach and AdaBoost methodology. Multim. Tools Appl. 80(7), 11155–11175 (2021)
15. Kingma, D.P., Ba, J.: Adam: A method for stochastic optimization. In: 3rd International Conference on Learning Representations, ICLR 2015, San Diego, CA, USA, May 7–9, 2015, Conference Track Proceedings (2015)
16. Le, H., Pino, J.M., Wang, C., Gu, J., Schwab, D., Besacier, L.: Lightweight adapter tuning for multilingual speech translation. In: Proceedings of the 59th Annual Meeting of the Association for Computational Linguistics and the 11th International Joint Conference on Natural Language Processing, ACL/IJCNLP 2021, (Volume 2: Short Papers), Virtual Event, 1–6 August 2021, pp. 817–824. Association for Computational Linguistics (2021)
17. Lewis, M., et al.: BART: denoising sequence-to-sequence pre-training for natural language generation, translation, and comprehension. In: Proceedings of the 58th Annual Meeting of the Association for Computational Linguistics, ACL 2020, Online, 5–10 July 2020, pp. 7871–7880. Association for Computational Linguistics (2020)
18. Li, X.L., Liang, P.: Prefix-tuning: optimizing continuous prompts for generation. In: Proceedings of the 59th Annual Meeting of the Association for Computational Linguistics and the 11th International Joint Conference on Natural Language Processing, ACL/IJCNLP 2021, (Volume 1: Long Papers), Virtual Event, 1–6 August 2021, pp. 4582–4597. Association for Computational Linguistics (2021)
19. Ma, C., et al.: Improving end-to-end text image translation from the auxiliary text translation task. In: 26th International Conference on Pattern Recognition, ICPR 2022, Montreal, QC, Canada, 21–25 August 2022, pp. 1664–1670. IEEE (2022)
20. Mansimov, E., Stern, M., Chen, M., Firat, O., Uszkoreit, J., Jain, P.: Towards end-to-end in-image neural machine translation. In: Proceedings of the First International Workshop on Natural Language Processing Beyond Text. Association for Computational Linguistics, November 2020
21. Papineni, K., Roukos, S., Ward, T., Zhu, W.: BLEU: a method for automatic evaluation of machine translation. In: Proceedings of the 40th Annual Meeting of the Association for Computational Linguistics, 6–12 July 2002, Philadelphia, PA, USA, pp. 311–318 (2002)
22. Radford, A., Narasimhan, K.: Improving language understanding by generative pre-training. Open AI Blog (2018)
23. Raffel, C., et al.: Exploring the limits of transfer learning with a unified text-to-text transformer. J. Mach. Learn. Res. 21, 140:1-140:67 (2020)
24. Rebuffi, S., Bilen, H., Vedaldi, A.: Learning multiple visual domains with residual adapters. In: Advances in Neural Information Processing Systems 30: Annual Conference on Neural Information Processing Systems 2017, 4–9 December 2017, Long Beach, CA, USA, pp. 506–516 (2017)

25. Rothe, S., Narayan, S., Severyn, A.: Leveraging pre-trained checkpoints for sequence generation tasks. Trans. Assoc. Comput. Linguist. **8**, 264–280 (2020)
26. Shekar, K.C., Cross, M.A., Vasudevan, V.: Optical character recognition and neural machine translation using deep learning techniques. In: Saini, H.S., Sayal, R., Govardhan, A., Buyya, R. (eds.) Innovations in Computer Science and Engineering. LNNS, vol. 171, pp. 277–283. Springer, Singapore (2021). https://doi.org/10.1007/978-981-33-4543-0_30
27. Shi, B., Bai, X., Yao, C.: An end-to-end trainable neural network for image-based sequence recognition and its application to scene text recognition. IEEE Trans. Pattern Anal. Mach. Intell. **39**(11), 2298–2304 (2017)
28. Shi, B., Wang, X., Lyu, P., Yao, C., Bai, X.: Robust scene text recognition with automatic rectification. In: 2016 IEEE Conference on Computer Vision and Pattern Recognition, CVPR 2016, Las Vegas, NV, USA, 27–30 June 2016, pp. 4168–4176. IEEE Computer Society (2016)
29. Su, T., Liu, S., Zhou, S.: RTNet: an end-to-end method for handwritten text image translation. In: Lladós, J., Lopresti, D., Uchida, S. (eds.) ICDAR 2021. LNCS, vol. 12822, pp. 99–113. Springer, Cham (2021). https://doi.org/10.1007/978-3-030-86331-9_7
30. Sun, Z., Wang, M., Li, L.: Multilingual translation via grafting pre-trained language models. In: Findings of the Association for Computational Linguistics: EMNLP 2021, Virtual Event/Punta Cana, Dominican Republic, 16–20 November 2021. ,p. 2735–2747. Association for Computational Linguistics (2021)
31. Sutskever, I., Vinyals, O., Le, Q.V.: Sequence to sequence learning with neural networks. In: Advances in Neural Information Processing Systems 27: Annual Conference on Neural Information Processing Systems 2014, 8–13 December 2014, Montreal, Quebec, Canada, pp. 3104–3112 (2014)
32. Vaswani, A., et al.: Attention is all you need. In: Advances in Neural Information Processing Systems 30: Annual Conference on Neural Information Processing Systems 2017, 4–9 December 2017, Long Beach, CA, USA, pp. 5998–6008 (2017)
33. Zaken, E.B., Goldberg, Y., Ravfogel, S.: BitFit: simple parameter-efficient fine-tuning for transformer-based masked language-models. In: Proceedings of the 60th Annual Meeting of the Association for Computational Linguistics (Volume 2: Short Papers), ACL 2022, Dublin, Ireland, 22–27 May 2022, pp. 1–9. Association for Computational Linguistics (2022)
34. Zhang, H., Li, G., Li, J., Zhang, Z., Zhu, Y., Jin, Z.: Fine-tuning pre-trained language models effectively by optimizing subnetworks adaptively. CoRR abs/2211.01642 (2022)
35. Zhang, Y., Nie, S., Liang, S., Liu, W.: Bidirectional adversarial domain adaptation with semantic consistency. In: Lin, Z., et al. (eds.) PRCV 2019. LNCS, vol. 11859, pp. 184–198. Springer, Cham (2019). https://doi.org/10.1007/978-3-030-31726-3_16
36. Zhang, Y., Nie, S., Liang, S., Liu, W.: Robust text image recognition via adversarial sequence-to-sequence domain adaptation. IEEE Trans. Image Process. **30**, 3922–3933 (2021)

37. Zhao, Y., Xiang, L., Zhu, J., Zhang, J., Zhou, Y., Zong, C.: Knowledge graph enhanced neural machine translation via multi-task learning on sub-entity granularity. In: Proceedings of the 28th International Conference on Computational Linguistics, COLING 2020, Barcelona, Spain (Online), 8–13 December 2020, pp. 4495–4505. International Committee on Computational Linguistics (2020)
38. Zhao, Y., Zhang, J., Zhou, Y., Zong, C.: Knowledge graphs enhanced neural machine translation. In: Proceedings of the Twenty-Ninth International Joint Conference on Artificial Intelligence, IJCAI 2020, pp. 4039–4045. ijcai.org (2020)

Open-Set Text Recognition via Shape-Awareness Visual Reconstruction

Chang Liu[1], Chun Yang[1], and Xu-Cheng Yin[1,2(✉)]

[1] School of Computer and Communication Engineering, University of Science and Technology Beijing, Beijing, China
lasercat@gmx.us, {chunyang,xuchengyin}@ustb.edu.cn
[2] Institute of Artificial Intelligence, University of Science and Technology Beijing, Beijing, China

Abstract. Open-Set Text Recognition (OSTR) is an emerging task that models the constantly evolving char-set in open-world character recognition applications. Compared to conventional counterparts, the OSTR task demands actively spotting and incrementally recognizing novel characters. Existing methods have demonstrated some success, yet confusion among similar characters remains to be a major challenge, potentially due to insufficient shape information preserved in the character features. In this work, we propose to alleviate this problem via visual reconstruction. Specifically, a glyph reconstruction task is adopted to implement shape awareness. Furthermore, cut-and-mixed characters are introduced to alleviate overfitting, by improving the coverage of the glyph space. Finally, a cycle classification task is proposed to prioritize the preservation of classification-critic regions by sending reconstructed images to a classifier network. Extensive experiments show that both tasks yield satisfying improvements on the OSTR task, and the full model demonstrates decent performance in recognizing both seen and novel characters.

Keywords: Open-set text recognition · Visual reconstruction · Low-shot learning

1 Introduction

Text recognition is a widely applied task, which is actively studied by both the research and industry communities [8]. However, the conventional close-set text recognition task fails to model several challenges of the open environment in real-world applications, which, in brief, are the linguistic model skew [41] and the occurrence of novel characters [25,26]. Specifically, the linguistic model skew refers to the out-of-vocabulary (OOV) samples in the testing set. Though the recognition capability of such samples dates back to decades ago [12], many modern methods are still vulnerable to the language model gap between the training data and the testing data [41].

G. A. Fink et al. (Eds.): ICDAR 2023, LNCS 14192, pp. 89–105, 2023.
https://doi.org/10.1007/978-3-031-41731-3_6

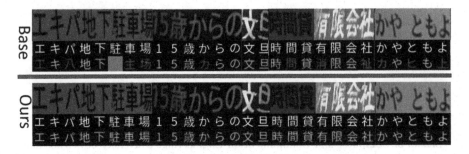

Fig. 1. Models lacking shape-awareness tends to confuse over visually close characters. For character colors, white: seen, yellow: novel, green: success, red: failure, and purple block indicates rejection. (Color figure online)

Novel characters refer to the unperceived characters that are not covered by the training set, which could be the results of the language evolution, the limitation on prior knowledge, or data availability issues. The emergence of novel characters demands two extra capabilities from the model, i.e., the spotting capability and the incremental recognition capability of such characters [25]. Here, the spotting capability demands the model to notify the users when unseen characters appear in the data stream, instead of recognizing them as seen characters in silence. The spotting capability also helps the user to take action to adjust the model timely to handle these novel characters. The incremental learning capability [54], on the other hand, helps avoid expensive data-collection and retraining processes during the adaption process.

To better model these challenges of the open-world environment, researchers propose the open-set text recognition (OSTR) task [25] as an extension to the generalized zero-shot text recognition tasks. The OSTR task aims to model both spotting and recognition of novel characters, without ignoring the performances on close-set benchmarks [3]. Existing methods like [25,26] have achieved the aforementioned goals to different extents, however, lack sufficient awareness of detailed character shape characteristics, demonstrating a tendency to confuse visually similar characters [26] (Also see Fig. 1). Due to the strong contextual information bias in open environments, alleviating this problem via language models is especially difficult.

In this work, we purpose an OpenSAVR network, which learns to preserve structural information in character features via auxiliary tasks. Specifically, we first adopt a glyph reconstruction task, which reconstructs generated prototypes (character class-centers) back to corresponding glyph images. To further improve the generalization capability of shape information preservation, we propose to augment the reconstruction task with cut-and-mixed glyphs to provide better coverage of the glyph space. In addition, we propose a cycle classification task to prioritize the preservation of discriminative visual features, by feeding reconstructed character images back to a character classifier. Since the character extracted from the crop-word samples are also extracted with the shared extractor and also aligned to the prototypes during training, the preservation of detailed shape information also applies.

Ablative experiments validate the effectiveness of each proposed strategy in improving recognition performance, and the OpenSAVR network demonstrates decent performances on both close-set [3] and open-world scenarios [25]. The appendixes, code, models, and documentation are released on Github[1]. In summary, our main contributions are:

– Introducing the glyph reconstruction auxiliary task to open-set text recognition methods.
– Propose to improve the glyph space coverage by reconstructing cut-and-mixed glyphs.
– Propose a cycle classification task to prioritize the preservation of more discriminative regions of each character.

2 Related Works

2.1 Text Recognition

Text recognition is a popular research topic widely applied in the industry, and is also upstream to various other tasks. Most text recognition methods focus on recognizing seen characters covered by the training set, and the performances are measured by the de facto standard protocol summarily formulated by Beak. et. al. [3]. In the early days, text recognition is implemented as a word classification task. To recognize out-of-vocabulary contents [12], methods break granularity down from words into more fine-grained components. Specifically, the granularity evolves from word level to gram level [23] and further to the current character-level recognition [24,36], which can be further categorized into feature aggregation methods and label aggregation methods.

Specifically, the label-aggregation methods [24,36,52] first produce position-wise classification results with regard to the 1D [36,52] or 2D [24] feature map extracted by the backbone, then aggregate the dense predictions into the final results. Commonly used aggregation algorithms include beam-search [36], connected components [24,40], and rule-based methods [52].

On the other hand, the feature aggregation methods [11,35,37] first aggregate extracted feature maps into character features, each corresponding to a character at a timestamp. A linear classifier is then used to decode individual character features into corresponding character label predictions. Earlier methods in this category produce each character feature according to its preceding timestamps [3,35,37]. Some more recent methods [11,44,50] relax the order constraint by parallel indexing with kernels [44] or quires [11,50]. Alternatively, some other methods utilize transformer layers to align features to time stamps in a latent manner [2].

---

[1] https://github.com/lancercat/OpenSAVR.

2.2   Open-Set Text Recognition

Real-life applications, especially those handle internet-oriented images, street signs, or historical documents are frequently challenged by Out-of-vocabulary (OOV) words [12] and new characters not covered by the training set. The new characters introduce the open-set text recognition task [25], which models two new challenges besides the existing recognition tasks [3,12], namely the spotting and incremental recognition capability over novel characters. The first challenge aims to spot unseen characters by predicting them into "unknown" labels, rather than making silent recognition errors, which falls in the larger scope of open-set recognition [13] and anomaly detection [29]. The second task demands incrementally recognizing new characters with provided side information, which falls in the scope of generalized zero-shot recognition [47] and incremental learning [54]. Overall, the OSOCR task is similar to the open-world object detection [15] and open-set segmentation tasks [19], however stressing more on the sequence nature of the label.

In the OSOCR task [25], characters are split into four categories, namely SIC (Seen In-set Characters), SOC(Seen Out-of-set Characters), NIC (Novel In-set Characters), and NOC (Novel Out-of-set Characters). Here, the first alphabet indicates whether the characters are covered by the training samples (Seen) or not (Novel). The second alphabet describes the availability of side-information during test time, splitting characters into "In-set" characters $\mathbf{C}_{test}^k$ and "Out-of-set" characters $\mathbf{C}_{test}^u$. Here, "in-set" characters, which include SICs and NICs, are provided with side information and are subject to recognition. The "out-of-set" characters denote any remaining characters not included in $\mathbf{C}_{test}^k$ that do not possess side information and are subject to rejection. Characters in $\mathbf{C}_{test}^u$ include two parts, the NOCs which are the unknown characters in the wild, and SOCs are not-interested characters or rare characters explicitly swapped out for speed up. To incrementally adapt to a seen or novel out-of-set character, the user can put it into $\mathbf{C}_{test}^k$ by providing its side information. Vise versa, the user can stop recognizing a character by removing its information, consequentially moving it from $\mathbf{C}_{test}^k$ to $\mathbf{C}_{test}^u$.

Currently, in the text recognition field, few methods [25] cover the spotting challenge, while most existing methods [18,26,52,53] aim to solve the recognition challenge. Among these methods, one category utilizes components like radical [18,53], stroke [6], or other compositional information [16] as side-information, while another category matches character features with visual knowledge like writing tracks [1] and glyphs [26,52] for recognition purposes. However, due to the compositional knowledge being mostly language-specific, many component-based methods have strong language dependencies. The visual-based methods are less language-dependent, however, some methods [26] demonstrate a tendency to confuse characters with similar shapes, presumably due to the lack of awareness of detailed components or structural clues. In this work, we propose to introduce visual reconstruction auxiliary tasks to alleviate this problem.

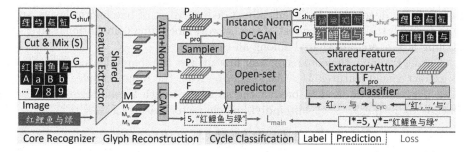

Fig. 2. The overall structure of the proposed OpenSAVR network. The network is composed of the Core Recognizer (In grey) and two proposed auxiliary co-training tasks, namely the Glyph Reconstruction task (in yellow) and the Cycle Classification task(in purple). During testing, the auxiliary tasks are removed and network reduces to the Core Recognizer, hence the proposed methods do not yield any extra inference cost.

2.3 Visual Reconstruction

Input reconstruction is a popular auxiliary task to improve feature representation in a variety of domains of deep learning [10,43,46]. The subject of reconstruction vary from features [17] to raw inputs (mainly text [10] or images [45,49]). The purposes also include, but are not limited to, out-of-distribution detection [17, 49], context modeling [10], and information preservation [31,43,45,46].

The information preserved could also be diverse, e.g., [31] reconstructs higher resolution images as an auxiliary task, while [27] only reconstructs "clean images" without background and style. EFIF [45] proposes to perform reconstruction on character visual features at each time stamp. Specifically, EFIF restyles the character feature with random font features and reconstructs the styled features back to the corresponding glyphs.

However, most of these methods are limited to reconstructing seen characters, which results in a limited number of distinctive classes. In addition, most methods treat different regions of the reconstructed image equally, which may cause the model to overlook minor, yet discriminative structural information or components. To address the limitation on training class numbers, we propose to augment existing labels into new classes via a cut-and-mix approach. A Cycle Classification task is also proposed to emphasize the preservation of more discriminative visual information in the auxiliary reconstruction task.

3 Our Method

To implement shape awareness of character features, we proposed the Open-SAVR network illustrated in Fig. 2. The network includes a core recognizer for inference and two training-time auxiliary tasks, i. e. the glyph reconstruction task (Sect. 3.1) and the cycle classification task (Sect. 3.2). The network utilizes a Resnet45-DSBN [26] network as a partially-shared feature extraction backbone for all tasks, denoted as T. Specifically, a dedicated set of batchnorm layers are used for each task, while the rest of the network, including convolutional layers, pooling layers, etc. are shared among all tasks.

During training, the core recognizer first utilizes the shared feature extractor to extract visual features $M : (M_h, M_m, M_l)$ from the input sample. Then the LCAM [26] module is used to predict the sequence length t_m and aggregate the feature map $M_l \in \mathbb{R}^{w \times h \times d}$ into individual character features $F \in \mathbb{R}^{t_m \times d}$,

$$\mathbf{Pr}(\hat{l}), F = \mathtt{LCAM}(M_h, M_m, M_l),$$
$$t_m = \mathtt{argmax}(\mathbf{Pr}(\hat{l})). \tag{1}$$

Next, the model samples a batch of c labels from the training label set $\mathbf{C}_{train}$ and generates corresponding n prototypes[2] $P \in \mathbb{R}^{n \times d}$ by feeding glyphs $G \in \mathbb{R}^{n \times 32 \times 32}$ to the prototype generator E, which consists of the shared feature extractor T and a spatial attention module Attn,

$$P = \mathtt{E}(G) := \mathtt{Norm}(\mathtt{Attn}(\mathtt{T}(G)))). \tag{2}$$

Specifically, the $\mathtt{Attn} : \mathbb{R}^{n \times w \times h \times d} \to \mathbb{R}^{n \times d}$ module reduces glyph feature maps to unnormalized prototypes [26][3], which then normalized via the Norm function implemented as a row-wise L2-Normalization.

The open-set predictor then decodes each character feature $F_{[t]}$ into the corresponding output label $\hat{y}_{[t]}$ by comparing $F_{[t]}$ to the generated prototypes P,

$$\mathbf{Pr}(\hat{y}_{[t]}) = \delta(|F_{[t]}|[\cos(F_{[t]}, P_{[1]}), ..., \cos(F_{[t]}, P_{[n]}), s_-]). \tag{3}$$

Here, $\mathbf{Pr}(\hat{y}_{[t]})$ is the predicted probability distribution of the character at timestamp t, [] is the concatenation operation, δ is the softmax function, and s_- is a shared radius of the decision boundaries of all classes. Character features not similar to any prototypes, which means yielding lower similarity scores than s_- with any of the prototypes in P, will be predicted as the "unknown" label.

The core recognition task is co-trained with two proposed auxiliary tasks aimed to improve the shape-awareness of prototypes P, which consequentially improves the character features F aligned to P. Specifically, the glyph reconstruction task is first utilized to ensure each prototype $P_{[i]}$ preserves all visual information of its corresponding glyph $G_{[i]}$, via reconstruction G from P. The glyph reconstruction task is further augmented with cut-and-mixed glyphs to achieve better coverage of the glyph space. Finally, the cycle classification task is utilized to further improve awareness of discriminative visual traits by classifying reconstructed characters back into labels.

During the evaluation, the network reduces to the core recognizer, and samples contain one or more "unknown" labels as prediction will be rejected. Furthermore, the prototypes can be cached and incrementally adjusted on characterset changes. As a result, the network only yields negligible extra computation cost during evaluation, compared to close-set text-recognition methods [44].

---

[2] Note a label may have more than one cases, and each case yields a dedicated prototype, thus $c \leq n$.

[3] Also described in Appendix A.2.

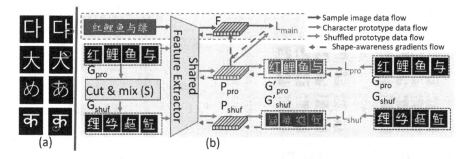

Fig. 3. (a) Example characters require detailed shape information to distinguish. (b) The overall structure of the glyph reconstruction task. The shared feature extractor is trained to preserve sufficient detailed shape information to reconstruct prototypes back to the corresponding glyphs.

3.1 Glyph Reconstruction

Detailed shape information is vital to distinguish visually close characters illustrated in Fig. 3 (a), especially when contextual information is intractable in open-world environments. Hence, it would be necessary for character features to preserve detailed shape information, and this trait is denoted as shape-awareness for short. In this work, we propose the glyph reconstruction auxiliary task shown in Fig. 3 (b) to implement shape awareness.

The task contains two parts yielding two corresponding losses, namely L_{pro} and L_{shuf}. Loss L_{pro} aims to make the generated prototypes include visual information needed to restore the corresponding glyphs. For each training iteration, the task first randomly samples n_{pro} prototypes P_{pro} from P, and their corresponding glyphs are denoted as G_{pro}. The module then reconstructs G'_{pro} from P_{pro} with the reconstruction function $\mathtt{C}$,

$$G'_{pro} = \mathtt{C}(P_{pro}) := \tanh(\alpha\mathtt{GAN}(P_{pro})). \tag{4}$$

Here, $\mathtt{GAN}$ indicates an adapted DC-GAN [32] generator with the original Batch Normalization layers replaced with Instance Normalization layers, to achieve better multi-domain training stability [21]. α is a constant value empirically set to 1.5 to alleviate the gradient vanish problem when the $\mathtt{tanh}$ function saturates.

The reconstructed glyphs are aligned to the sampled glyphs with a balanced L2-loss $\mathtt{L_{bl2}}$,

$$
\begin{aligned}
L_{pro} &= \mathtt{L_{bl2}}(G'_{pro}, G_{pro}) \\
&= \frac{\sum_{I,j}^{w,h}(G'_{pro} - G_{pro})^2 G_{pro}}{2\sum G_{pro}} + \frac{\sum(G'_{pro} - G_{pro})^2(1 - G_{pro})}{2\sum(1 - G_{pro})}.
\end{aligned} \tag{5}
$$

Albeit the trained DC-GAN is able to reconstruct seen prototypes pretty well, we found it fails to reconstruct novel prototypes during testing, which means

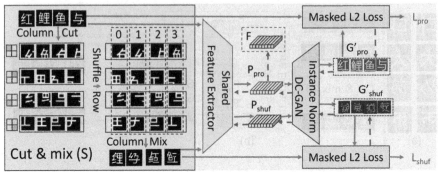

Fig. 4. The detailed process of the cut-and-mix glyph augmentation. The module breaks each character into parts and then re-combines the parts randomly to create new characters.

the shape awareness does not properly hold for the whole glyph space. To alleviate this problem, we propose to augment the glyphs for better coverage, by introducing the cut-and-mix reconstruction loss L_{shuf}.

A cut-and-mix module S (shown in Fig. 4) is first proposed to generate synthetic unseen glyphs. Specifically, it first cuts each glyph into four equal regions. Next, all parts in each individual region are shuffled and re-matched to generate new combinations, which are then reassembled into the synthetic glyphs G_{shuf}, Given the synthetic glyphs, the module generates corresponding prototypes P_{shuf} with the prototype generator module E defined in Eq. 2), and reconstructs them back to glyph images G'_{shuf} with the reconstructor C,

$$G_{shuf} = \text{S}(G), G'_{shuf} = \text{C}(\text{E}(G_{shuf})). \tag{6}$$

Similarly, a balanced L2 loss is used to align G'_{shuf} and G_{shuf},

$$L_{shuf} = \text{L}_{bl2}(G'_{shuf}, G_{shuf}). \tag{7}$$

In summary, the glyph reconstruction task produces the overall reconstruction loss L_{rec} as the summation of L_{pro} and L_{shuf},

$$L_{rec} = L_{pro} + L_{shuf}. \tag{8}$$

Since the character features F are aligned to prototypes, this task also helps to improve the shape-awareness of F.

3.2   Cycle Classification

Though direct supervising is viable for reconstructing prototypes, it treats discriminative and non-discriminative regions equally, which may lead the model to neglect small but important parts. Thus, in this work, we propose a cycle classification task that further re-classifies the character images G'_{pro} reconstructed from prototypes P back to labels, yielding the cycle loss L_{cyc}. Here, the shuffled glyphs G'_{shuf} are omitted to reduce training variance. Specifically, the shared feature extractor is first used to extract features from the reconstructed glyph images,

$$F_{pro} = \texttt{Attn}(\texttt{T}(G'_{pro})) \tag{9}$$

Then the features F_{pro} are sent to a classifier, yielding the cycle loss L_{cyc},

$$\mathbf{Pr}(\boldsymbol{y}_{cyc}) = \delta(|F_{pro}|[\texttt{cos}(F_{pro}, P_{[1]}), ..., \texttt{cos}(F_{pro}, P_{[n_{pro}]})]) \tag{10}$$

Here, n_{pro} is the total number of the reconstructed character prototypes, $\boldsymbol{y}^*_{cyc[i]}$ are the corresponding labels in the one-hot form, and P is the corresponding prototypes in the batch, described in Sect. 3.

3.3 Optimization

The training of the core recognizer yields the recognition loss L_{main}, which is the combination of the cross-entropy losses on sample length prediction and label prediction,

$$L_{main} = \texttt{log}(\mathbf{Pr}(\hat{l})^T l^*) + \sum_{t=1}^{t^*_m} \texttt{log}(\mathbf{Pr}(\hat{\boldsymbol{y}}_{[t]})^T \boldsymbol{y}^*_{[t]}), \tag{11}$$

where l^* and $\boldsymbol{y}^*_{[t]}$ are one-hot representations of the ground truth of sequence length (t^*_m) and the character label at timestamp t. The model can hence be trained by optimizing the recognition loss L_{main} together with the losses of the auxiliary tasks,

$$L = L_{main} + \lambda_{rec}L_{rec} + \lambda_{cyc}L_{cyc}. \tag{12}$$

Here λ_{rec} is the weight of the glyph reconstruction task and λ_{cyc} is the weight of the cycle classification task, which are both set to 0.3 empirically.

4   Experiments

In this work, we first perform ablative studies to validate each proposed module's performance on the OSOCR dataset [25]. As the modules aim to improve recognition performance, we use the GZSL protocol [25] which focuses on recognition for ablative studies. The full model is then evaluated with various open-set text recognition protocols to validate its capabilities for spotting and incrementally recognizing novel characters, as well as the recognition performance on seen characters. Finally, we evaluate the full model on the standard close-set text recognition benchmark [3] to validate its feasibility to replace lightweight conventional models in close-set environments.

4.1 Implementation Details

The method is derived from the OpenCCD [26] codebase, which is built on the Pytorch framework. Regular models can be trained on Nvidia GPUs with 8 GiB of Vram. Besides the regular model, we also provide a large model (Ours-large) which has 1.5x channels and needs 24 GiB of Vram to train. The model is trained on the same training data as the regular model and is also included in the released code. Both regular and large models have a low evaluation profile and can be evaluated with only 2 GiB of Vram. Models have trained for 2×10^5 iterations for open-set experiments and 10^6 iterations for the close-set benchmarks due to the larger training set. To reduce confounding factors, all ablative experiments are conducted on the same server with 4 1080Tis. The torch version on that server is 1.7.1 and torchvision is at 0.8.2. Evaluation is conducted on a workstation with one Tesla P40 GPU and 2 E5-2620v3 CPUs.

4.2 Bench-Marking Protocols

This paper adopts two benchmarks, namely the Open-set Text recognition [25] and the standard close-set text recognition benchmark [3]. Metric-wise, the recognition accuracy is measured by line accuracy (LA) [3,25],

$$LA = \frac{\sum_{i=1}^{N} \texttt{same}(GT_i, PR_i)}{N}, \tag{13}$$

where N is the number of word samples without out-of-set characters in the testing dataset. For close-set benchmarks and the GZSL split, N equals the total number of samples in the testing set. Other OSTR splits also measure the novel-character-spotting performance, via word-level recall (RE), precision (PR), and F-measure (FM). Recall and Precision are computed as follows,

$$RE = \frac{\sum_i^N \texttt{Rej}(PR_{[I]})\texttt{Rej}(GT_{[I]})}{\sum_i^N \texttt{Rej}(GT_{[I]})}, \ PR = \frac{\sum_i^N \texttt{Rej}(PR_{[I]})\texttt{Rej}(GT_{[I]})}{\sum_i^N \texttt{Rej}(PR_{[I]})}, \tag{14}$$

where $\texttt{Rej}$ returns whether the string contains characters not in $\mathbf{C}_{test}^k$, or predicted unknown label. The F-measure is the harmonic mean of recall and precision, defined as

$$FM = \frac{2RE * PR}{RE + PR}. \tag{15}$$

Dataset-wise, open-set models are trained on Chinese and English samples collected from a variety of public datasets, including the ART [9], RCTW [38], LSVT [39], CTW [51], and the Latin-Chinese subset of the MLT dataset. Training samples are filtered with 3755 tire-1 simplified Chinese characters, 26 English letters, and digits, where samples including other characters are excluded from the training set to prevent label leakage. The model is then evaluated on the Japanese testing samples collected from the MLT Japanese subset under 4 different character splits to model different use stages and preferences in the use

Table 1. Ablative results and sensitivity analysis on the GZSL setup.

Name	L_{pro}	L_{shuf}	L_{cyc}	$\lambda_{cyc}\&\lambda_{rec}$	Line Accuracy
Base				-	36.89
L_{pro}	✓			0.3	38.89
$L_{pro}+L_{shuf}$	✓	✓		0.3	40.71
Full model	✓	✓	✓	0.3	**40.96**
Full model-0.1	✓	✓	✓	0.1	37.31
Full model-0.5	✓	✓	✓	0.5	40.38

cases illustrated in Appendix B.2, namely the GZSL split, the OSR split, the GOSR split, and the OSTR split. Specifically, the GZSL split reduces to the Generalized Zero-shot Learning which focuses on measuring recognition capability. The OSR split reduces to the Open-Set Recognition, which focuses on spotting novel characters. The GOSR split is close to the Generalized Open-set Recognition [13], which handles recognition and spotting at the same time. Finally, the OSTR split includes seen characters as subjects of rejection. Note unlike common OSR tasks [13], the full OSTR task includes the rejection of seen characters, which correspond to the use-case where infrequent characters are temporarily removed to speed up the classification process. Samples with such characters will first trigger a rejection, where the user can load back previously removed prototypes for recognition.

For close-set experiments, we train the model on the MJ [20] and ST [14] synthetic datasets, following the standard benchmarking protocol [3]. The performance is evaluated on the IIIT5k [30], CUTE [33], SVT [42], IC03 [28], and IC13 [22] datasets. Similar to [3], we measure the model performance accuracy-wise via Line Accuracy and speed-wise via single batched inference speed.

4.3   Ablative Studies and Sensitivity Analysis

The ablative studies are conducted on the GZSL setup of the open-set text recognition task, which focuses on the recognition performance of seen and novel characters, and the results are shown in Table 1. The results indicate that all proposed approaches, i.e., the glyph reconstruction, the cut-and-mix label augmentation, and the cycle classification module can improve the recognition performance of novel language. Specifically, we find reconstructing glyphs from prototypes with L_{pro} improves the Line Accuracy by 2 %, while augmenting glyph reconstruction via cut-and-mix (L_{shuf}) yields another 1.9 % of improvement. Focusing on discriminative parts with L_{cyc} further contributes a small performance gain of 0.25 percent. Thus, we consider the proposed approaches valid in improving open-set text recognition performance. Qualitative samples shown in Fig. 1 also suggest the full model demonstrates a better capability to differentiate characters with similar shapes compared to the baseline model.

Table 2. Performance on the open-set text recognition task. Bold indicate main metrics, and italic indicates characters unseen during training. $\mathbf{C}^k_{test}$ indicates characters with side information subjects to recognition, while $\mathbf{C}^u_{test}$ indicates characters without side information subjects to rejection.

Split	$\mathbf{C}^k_{test}$	$\mathbf{C}^u_{test}$	Name	LA	Recall	Precision	F-measure
GZSL	*Unique Kanji*	∅	OSOCR-Large [25]	30.83	–	–	–
	Shared Kanji,		OpenCCD [26]	36.57	–	–	–
	Kana, Latin,		OpenCCD-Large [26]	41.31	–	–	–
			Ours	40.96	–	–	–
			Ours-Large	**42.58**	–	–	–
OSR	Shared Kanji	*Unique*	OSOCR-Large [25]	74.35	11.27	**98.28**	20.23
	Latin	*Kanji*	**Ours**	75.08	56.38	96.95	71.29
		Kana	**Ours-Large**	**78.49**	**58.81**	97.33	**73.32**
GOSR	Shared Kanji	*Kana*	OSOCR-Large [25]	56.03	3.03	63.52	5.78
	Unique Kanji		**Ours**	68.43	34.23	80.58	48.05
	Latin		**Ours-Large**	**69.29**	**41.19**	**85.05**	**55.50**
OSTR	Shared Kanji,	*Kana*	OSOCR-Large [25]	58.57	24.46	**93.78**	38.80
	Unique Kanji	Latin	**Ours**	71.86	69.72	90.86	78.90
			Ours-Large	**72.33**	**72.96**	92.62	**81.62**

In this work, we set the weights (λ_{cyc} and λ_{rec}) of the proposed losses to 0.3 as an approximation to $\sqrt{\frac{1}{10}}$, which is neither too large nor small compared to the classification loss. We also conducted sensitivity analysis over the weights, and the full model always shows a decent performance advantage against the base model. Specifically, we also set the value to 0.1 and 0.5 as shown in Table 1. Although the performances are always better than the base model, the improvement is less significant with too large or too small weights. Specifically, setting a larger weight reduces the performance gain by 0.6 percent, which indicates that the proposed losses can destabilize the training process, but does yield significant adversary effects as the performance is still 3.5 percent higher than the base model. Reducing the weight to 0.1, which limits the distillation speed of the shape-awareness prior, consequentially yields positive but limited improvement against the base model, which indicates that shape-awareness is necessary for the model to generalize to open environments.

4.4 Openset Performance

We mainly follow the protocols in OSOCR [25] to evaluate the capabilities to spot (reject) novel characters, and incrementally learn to recognize novel characters, without forgetting seen characters.

Quantitative results are shown in Table 2, and the performance is generally better than existing methods [25,26]. Furthermore, the capability to spot novel characters in the datastream also reaches a practical level. Qualitatively, we illustrate representative recognition results of the GZSL split in Fig. 5. Results indicate the model is generally robust with clear images, while still vulnerable

Fig. 5. Qualitative samples of Ours-Large. Results are categorized by the character types in the samples. For character colors, white: seen, yellow: novel, green: success, red: failure, and purple block indicates rejection. (Color figure online)

Table 3. Accuracy and single-batched speed comparisons on close-set benchmarks

Method	Venue	IIIT5K	CUTE	SVT	IC03	IC13	GPU	TFlops	FPS
Rosetta [3,5]	KDD'18	84.3	69.2	84.7	92.9	89.0	Tesla P40	12	**212**
Comb.Best [3]	ICCV'19	87.9	74.0	87.5	94.4	92.3	Tesla P40	12	36
JVSR [4]	ICCV'21	**95.2**	**89.7**	**92.2**	-	**95.5**	RTX 2080Ti	13.6	38
CA-FCN* [24]	AAAI'19	92.0	79.9	82.1	-	91.4	Titan XP	12	45
PERN [48]	CVPR'21	92.1	81.3	92.0	**94.9**	**94.7**	Tesla V100	14	44
ViTSTR [2]	ICDAR'21	88.4	81.3	87.7	94.3	92.4	RTX 2080Ti	13.6	102
Ours	-	90.67	73.26	82.53	87.54	89.26	Tesla P40	12	60
Ours-Large	-	92.33	84.38	85.16	91.58	90.64	Tesla P40	12	54

to blurs and other distortions, as a result of lacking modeling of linguistic information. However, modeling such information of evolved or unseen languages is still a challenging topic and is out of the scope of this work.

4.5 Close-Set Performance

Additionally, we evaluate the proposed method on standard close-set benchmarks [3]. Qualitative results are shown in Table 3, while quantitative results are shown in Fig. 6.

The regular model demonstrates comparable performance to recent lightweight methods at 60 FPS in single-batched evaluation [3]. The large model achieves decent performance, specifically within 3% below heavy SOTA methods [4] on the largest IIIT5k dataset, while still keeping a fast speed profile at 54FPS. Overall, the proposed model demonstrates decent feasibility for conventional applications as well as a lightweight method.

Fig. 6. Qualitative samples of Ours-Large. For character colors, white: gt, green: success, red: failure, and the purple block indicates rejection. (Color figure online)

5   Limitation

Despite demonstrating decent open-set recognition and rejection performance on the Japanese language, the model demonstrates less satisfying generalization capability against languages with less visual similarity to the training data, e.g., Koreans, Hindi, and Bengali, which forms the topic of our future projects.

Another limitation is that the close-set performance is only benchmarked with the English language [3], which might be insufficient to cover more complex multi-lingual use cases. Albeit there exist emerging benchmark protocols on various languages [7,34], they are yet to be well recognized by the community. Hence, we choose to wait till the next well-accepted standard emerges.

6   Conclusion

In summary, the ablative studies validate the effectiveness of the proposed glyph reconstruction module and cycle classification module as auxiliary training tasks, and the full model of OpenSAVR achieves overall satisfactory recognition and rejection performance on the open-set text recognition tasks. Meanwhile, the proposed OpenSAVR model suffices as a decent lightweight recognition method under the close-set environment, making it a feasible substitution for existing popular text recognition methods widely used in the industry [5,36].

Acknowledgement. The research is supported by National Key Research and Development Program of China (2020AAA0109700), National Science Fund for Distinguished Young Scholars (62125601), National Natural Science Foundation of China (62076024, 62006018), Interdisciplinary Research Project for Young Teachers of USTB (Fundamental Research Funds for the Central Universities)(FRF-IDRY-21-018).

References

1. Ao, X., Zhang, X., Yang, H., Yin, F., Liu, C.: Cross-modal prototype learning for zero-shot handwriting recognition. In: ICDAR, pp. 589–594 (2019)
2. Atienza, R.: Vision transformer for fast and efficient scene text recognition. In: Lladós, J., Lopresti, D., Uchida, S. (eds.) ICDAR 2021. LNCS, vol. 12821, pp. 319–334. Springer, Cham (2021). https://doi.org/10.1007/978-3-030-86549-8_21
3. Baek, J., et al.: What is wrong with scene text recognition model comparisons? dataset and model analysis. In: ICCV, pp. 4714–4722 (2019)
4. Bhunia, A.K., Sain, A., Kumar, A., Ghose, S., Chowdhury, P.N., Song, Y.: Joint visual semantic reasoning: Multi-stage decoder for text recognition. In: ICCV, pp. 14920–14929 (2021)
5. Borisyuk, F., Gordo, A., Sivakumar, V.: Rosetta: Large scale system for text detection and recognition in images. In: KDD, pp. 71–79 (2018)
6. Chen, J., Li, B., Xue, X.: Zero-shot Chinese character recognition with stroke-level decomposition. In: IJCAI, pp. 615–621 (2021)
7. Chen, J., et al.: Benchmarking Chinese text recognition: Datasets, baselines, and an empirical study. arXiv preprint arXiv:2112.15093 (2021)

8. Chen, X., Jin, L., Zhu, Y., Luo, C., Wang, T.: Text recognition in the wild: a surcvey. CSUR **54**(2), 1–35 (2021)
9. Chng, C.K., et al.: ICDAR2019 robust reading challenge on arbitrary-shaped text - rrc-art. In: ICDAR, pp. 1571–1576 (2019)
10. Devlin, J., Chang, M.W., Lee, K., Toutanova, K.: Bert: pre-training of deep bidirectional transformers for language understanding. arXiv preprint arXiv:1810.04805 (2018)
11. Fang, S., Xie, H., Wang, Y., Mao, Z., Zhang, Y.: Read like humans: autonomous, bidirectional and iterative language modeling for scene text recognition. In: CVPR, pp. 7098–7107 (2021)
12. Garcia-Bordils, S., et al.: Out-of-vocabulary challenge report. arXiv preprint arXiv:2209.06717 (2022)
13. Geng, C., Huang, S., Chen, S.: Recent advances in open set recognition: a survey. IEEE Trans. Pattern Anal. Mach. Intell. **43**(10), 3614–3631 (2021)
14. Gupta, A., Vedaldi, A., Zisserman, A.: Synthetic data for text localisation in natural images. In: CVPR, pp. 2315–2324 (2016)
15. Han, J., Ren, Y., Ding, J., Pan, X., Yan, K., Xia, G.S.: Expanding low-density latent regions for open-set object detection. In: Proceedings of the IEEE/CVF Conference on Computer Vision and Pattern Recognition, pp. 9591–9600 (2022)
16. He, S., Schomaker, L.: Open set Chinese character recognition using multi-typed attributes. arXiv preprint arXiv:1808.08993 (2018)
17. Huang, H., Wang, Y., Hu, Q., Cheng, M.M.: Class-specific semantic reconstruction for open set recognition. In: IEEE TPAMI (2022)
18. Huang, Y., Jin, L., Peng, D.: Zero-shot Chinese text recognition via matching class embedding. In: Lladós, J., Lopresti, D., Uchida, S. (eds.) ICDAR 2021. LNCS, vol. 12823, pp. 127–141. Springer, Cham (2021). https://doi.org/10.1007/978-3-030-86334-0_9
19. Hwang, J., Oh, S.W., Lee, J.Y., Han, B.: Exemplar-based open-set panoptic segmentation network. In: Proceedings of the IEEE/CVF Conference on Computer Vision and Pattern Recognition, pp. 1175–1184 (2021)
20. Jaderberg, M., Simonyan, K., Vedaldi, A., Zisserman, A.: Synthetic data and artificial neural networks for natural scene text recognition. CoRR abs/1406.2227 (2014)
21. Jin, X., Lan, C., Zeng, W., Chen, Z.: Style normalization and restitution for domain generalization and adaptation. IEEE Trans. Multimedia **24**, 3636–3651 (2021)
22. Karatzas, D., Shafait, F., Uchida, S., Iwamura, M., i Bigorda, L.G., Mestre, S.R., Mas, J., Mota, D.F., Almazán, J., de las Heras, L.: ICDAR 2013 robust reading competition. In: ICDAR. pp. 1484–1493 (2013)
23. Krishnan, P., Dutta, K., Jawahar, C.: Deep feature embedding for accurate recognition and retrieval of handwritten text. In: ICFHR, pp. 289–294. IEEE (2016)
24. Liao, M., et al.: Scene text recognition from two-dimensional perspective. In: AAAI, pp. 8714–8721 (2019)
25. Liu, C., Yang, C., Qin, H.B., Zhu, X., Liu, C.L., Yin, X.C.: Towards open-set text recognition via label-to-prototype learning. Pattern Recogn. **134**, 109109 (2023)
26. Liu, C., Yang, C., Yin, X.C.: Open-set text recognition via character-context decoupling. In: CVPR, pp. 4523–4532, June 2022
27. Liu, Y., Wang, Z., Jin, H., Wassell, I.: Synthetically supervised feature learning for scene text recognition. In: Ferrari, V., Hebert, M., Sminchisescu, C., Weiss, Y. (eds.) ECCV 2018. LNCS, vol. 11209, pp. 449–465. Springer, Cham (2018). https://doi.org/10.1007/978-3-030-01228-1_27
28. Lucas, S.M., et al.: ICDAR 2003 robust reading competitions: entries, results, and future directions. Int. J. Doc. Anal. Recognit. **7**(2–3), 105–122 (2005)

29. Ma, X., et al.: A comprehensive survey on graph anomaly detection with deep learning. arXiv preprint arXiv:2106.07178 (2021)
30. Mishra, A., Alahari, K., Jawahar, C.: Scene text recognition using higher order language priors. In: BMVC, BMVA (2012)
31. Mou, Y., et al.: PlugNet: degradation aware scene text recognition supervised by a pluggable super-resolution unit. In: Vedaldi, A., Bischof, H., Brox, T., Frahm, J.-M. (eds.) ECCV 2020. LNCS, vol. 12360, pp. 158–174. Springer, Cham (2020). https://doi.org/10.1007/978-3-030-58555-6_10
32. Radford, A., Metz, L., Chintala, S.: Deep convolutional generative adversarial network. In: Under Review as a Conference Paper at ICLR (2016)
33. Risnumawan, A., Shivakumara, P., Chan, C.S., Tan, C.L.: A robust arbitrary text detection system for natural scene images. Expert Syst. Appl. **41**(18), 8027–8048 (2014)
34. Shen, Z., Zhang, K., Dell, M.: A large dataset of historical Japanese documents with complex layouts. arXiv preprint arXiv:2004.08686 (2020)
35. Sheng, F., Chen, Z., Xu, B.: NRTR: a no-recurrence sequence-to-sequence model for scene text recognition. In: ICDAR, pp. 781–786 (2019)
36. Shi, B., Bai, X., Yao, C.: An end-to-end trainable neural network for image-based sequence recognition and its application to scene text recognition. IEEE Trans. Pattern Anal. Mach. Intell. **39**(11), 2298–2304 (2017)
37. Shi, B., Yang, M., Wang, X., Lyu, P., Yao, C., Bai, X.: ASTER: an attentional scene text recognizer with flexible rectification. IEEE Trans. Pattern Anal. Mach. Intell. **41**(9), 2035–2048 (2019)
38. Shi, B., et al.: ICDAR2017 competition on reading Chinese text in the wild (RCTW-17). In: ICDAR, pp. 1429–1434 (2017)
39. Sun, Y., et al.: ICDAR 2019 competition on large-scale street view text with partial labeling - RRC-LSVT. In: ICDAR, pp. 1557–1562 (2019)
40. Wan, Z., He, M., Chen, H., Bai, X., Yao, C.: Textscanner: reading characters in order for robust scene text recognition. In: AAAI, pp. 12120–12127 (2020)
41. Wan, Z., Zhang, J., Zhang, L., Luo, J., Yao, C.: On vocabulary reliance in scene text recognition. In: CVPR, pp. 11422–11431 (2020)
42. Wang, K., Babenko, B., Belongie, S.J.: End-to-end scene text recognition. In: ICCV, pp. 1457–1464 (2011)
43. Wang, L., Li, D., Zhu, Y., Tian, L., Shan, Y.: Dual super-resolution learning for semantic segmentation. In: CVPR, pp. 3773–3782 (2020)
44. Wang, T., et al.: Decoupled attention network for text recognition. In: AAAI, pp. 12216–12224 (2020)
45. Wang, Y., Lian, Z.: Exploring font-independent features for scene text recognition. In: MM, pp. 1900–1920 (2020)
46. Wang, Y., Lian, Z., Tang, Y., Xiao, J.: Boosting scene character recognition by learning canonical forms of glyphs. Int. J. Doc. Anal. Recogn. (IJDAR) **22**(3), 209–219 (2019). https://doi.org/10.1007/s10032-019-00326-z
47. Xian, Y., Lampert, C.H., Schiele, B., Akata, Z.: Zero-shot learning - a comprehensive evaluation of the good, the bad and the ugly. IEEE Trans. Pattern Anal. Mach. Intell. **41**(9), 2251–2265 (2019)
48. Yan, R., Peng, L., Xiao, S., Yao, G.: Primitive representation learning for scene text recognition. In: CVPR, pp. 284–293 (2021)
49. Yoshihashi, R., Shao, W., Kawakami, R., You, S., Iida, M., Naemura, T.: Classification-reconstruction learning for open-set recognition. In: CVPR, pp. 4016–4025 (2019)

50. Yu, D., et al.: Towards accurate scene text recognition with semantic reasoning networks. In: CVPR, pp. 12110–12119 (2020)
51. Yuan, T., Zhu, Z., Xu, K., Li, C., Mu, T., Hu, S.: A large Chinese text dataset in the wild. J. Comput. Sci. Technol. **34**(3), 509–521 (2019)
52. Zhang, C., Gupta, A., Zisserman, A.: Adaptive text recognition through visual matching. In: Vedaldi, A., Bischof, H., Brox, T., Frahm, J.-M. (eds.) ECCV 2020. LNCS, vol. 12361, pp. 51–67. Springer, Cham (2020). https://doi.org/10.1007/978-3-030-58517-4_4
53. Zhang, J., Du, J., Dai, L.: Radical analysis network for learning hierarchies of Chinese characters. Pattern Recognit. **103**, 107305 (2020)
54. Zhou, Z.H.: Open-environment machine learning. Nat. Sci. Rev. **9**(8), nwac123 (2022)

Accelerating Transformer-Based Scene Text Detection and Recognition via Token Pruning

Sergi Garcia-Bordils[1,2]([⊠]) [iD], Dimosthenis Karatzas[1] [iD], and Marçal Rusiñol[2] [iD]

[1] Computer Vision Center, Universitat Autonoma de Barcelona, Barcelona, Spain
{sergi.garcia,dimos}@cvc.uab.cat
[2] AllRead MLT, Barcelona, Spain

Abstract. Scene text detection and recognition is a crucial task in computer vision with numerous real-world applications. Transformer-based approaches are behind all current state-of-the-art models and have achieved excellent performance. However, the computational requirements of the transformer architecture makes training these methods slow and resource heavy. In this paper, we introduce a new token pruning strategy that significantly decreases training and inference times without sacrificing performance, striking a balance between accuracy and speed. We have applied this pruning technique to our own end-to-end transformer-based scene text understanding architecture. Our method uses a separate detection branch to guide the pruning of uninformative image features, which significantly reduces the number of tokens at the input of the transformer. Experimental results show how our network is able to obtain competitive results on multiple public benchmarks while running at significantly higher speeds.

Keywords: Scene Text Detection · Scene Text Recognition · Transformer Acceleration

1 Introduction

Joint text detection and recognition has become a popular topic in the field of computer vision for its wide range of applications. Text is omnipresent in man-made environments, and it plays a crucial role in different computer vision tasks such as visual-question answering [2, 45] or cross-modal retrieval [33], and in many computer vision applications like autonomous navigation [42] or industrial automation [8].

Early deep-learning based systems for text detection and recognition were based on two-stage pipelines, a detection network that extracted regions of interest (RoI) and a recognition network that recognized the cropped regions. The two tasks were treated as separate problems, with no gradient flowing between the two networks [24, 25]. More recent works attempted to jointly optimize both parts of the pipeline, allowing end-to-end trainable architectures [11, 23, 28, 32, 49]. A

G. A. Fink et al. (Eds.): ICDAR 2023, LNCS 14192, pp. 106–121, 2023.
https://doi.org/10.1007/978-3-031-41731-3_7

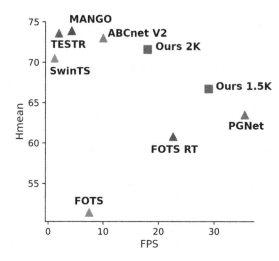

Fig. 1. Comparison between inference speed (in frames per second) and Hmean on ICDAR15 for different state-of-art scene text detection and recognition models. Our approach offers a balance between performance and inference speed thanks to our novel token pruning. The reported results use two different image scales (1500 and 2000).

common drawback of these networks is that they needed to explicitly rectify the RoI before they can be fed into the recognizer, which usually reads from left to right. For example, the authors of FOTS [27], a network that detects rotated bounding boxes, rectified the rotation of the RoI with their proposed RoIRotate operation. Other models, such as AbcNet [28] or Mask TextSpotter [23] proposed more complex de-warping techniques that were able to rectify heavily distorted text, such as curved text.

More recently, different one-stage methods have started appearing that do not require corrective operations on the detected areas [13,18,19,39,51]. Many of these approaches are based on the transformer architecture proposed by Vaswani et al. [47]. The common approach is to pass the features extracted by a CNN to the transformer, where the powerful self-attention mechanism performs detection and recognition. The fully connected topology of the self-attention removes the need to use RoI corrective operations. One of the drawbacks of the transformer is the quadratic complexity of the self-attention mechanism with respect to the number of input tokens. Increasing the input image resolution results in significantly slower training and inference times and higher memory usage.

The $O(n^2)$ complexity of the transformer has motivated multiple works that attempt to improve the efficiency of architecture. Some NLP approaches [10,50] have attempted modifying the fully connected self-attention with simpler topologies that reduce the complexity to $O(n)$. On vision and ViT-based [5] models, a popular approach is to reduce the number of tokens by employing a sampling/pruning mechanism to progressively discard uninformative tokens [6,22,35,41]. Many of these strategies such as ATS [6], or EViT [22] have been specially tailored for classification tasks on the ViT [5] architecture, and can not be directly applied to the object detection. Approaches like DynamicViT [41] or IA-RED^2 [35] need to train a specific component of the network to remove tokens, which often employs complex strategies such as reinforcement learning.

In this paper we introduce a novel token-pruning mechanism that has been specifically designed for scene text detection and recognition models. Our pruning approach works under the assumption that visual information of text is very local, while most of the background area is non-informative. The pruning mechanism reduces the complexity of the model and allows more efficient training parallelization and lower inference times. This pruning strategy has been applied to our own transformer encoder-based architecture, which is capable of reading text in multiple orientations, curvatures and distortions without needing to perform RoI corrections. Figure 2 shows an example of how our network performs pruning over the visual features before they go into the recognition branch. Our network has been designed to achieve a balance between quantitative performance and high inference speeds. As seen in Fig. 1, our model manages to get competitive results with the state-of-art while running at higher FPS. The contributions of this paper are the following:

- A novel token-pruning mechanism that allows the architecture to reduce the size of the input to the recognizer branch, a transformer-encoder network. We show how our strategy yields lower training and inference times than cropping and recognizing the detected areas independently.
- An efficient end-to-end text detection and recognition architecture where both tasks happen independently of each other. The model does not require any type of RoI corrective operations over the detected areas thanks to the fully connected attention mechanism.
- We show that our method manages to balance performance and speed, allowing us to reach competitive results with the state-of-the-art at significantly higher inference speeds.

2   Related Work

2.1   End-To-End Scene Text Recognition

Scene text detection and recognition is a challenging topic that has been an active area of research for many years. The complexity and sophistication of the architectures increased as different datasets and annotation styles started emerging. The different scenarios for text detection feature horizontal bounding boxes [17], incidental text with 4-point annotations [16,34,46,48] and, more recently, arbitrarily shaped text [3,29] such as curved text, which often feature complex polygonal annotations.

Earlier models, such as Textboxes [25], directly cropped the detected horizontal boxes from the image, and performed no rectification to the crop. This was a problem for the most commonly used recognition architectures, such as the encoder-decoder [21,21] and the CTC-based [9] networks, because they require

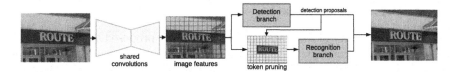

Fig. 2. The proposed architecture uses a shared convolutional backbone that extracts visual features from the image and then up-scales the output feature map. Two separate branches perform detection and recognition. We use the detection branch bounding boxes to guide the pruning mechanism, reducing the number of tokens at the input of the transformer. During training, we use the ground truth localization of the text to prune the image features.

the text to be horizontal and from left to right. As the complexity of annotations increased, models started to feature two-stage architectures that were fully end-to-end trainable and performed complex RoI rectifications to the detected areas.

More complex annotations and RoI corrective operations allowed recognition of text in different shapes and orientations. For example, FOTS [27] uses a text detection branch to predict oriented text boxes, and uses RoIRotation to obtain axis-aligned feature maps. The text recognition branch uses a bidirectional LSTM [12,44] and a CTC [9] network to recognize the crop. Mask TextSpotterV3 [23] extracts rectangular crops from the segmentation and then masks-out the area outside of the region of interest. ABCNet [28] uses a more unconventional approach by fitting Bezier curves to the text instances, which helps to obtain smoother boundaries around the words. The curves are rectified with their proposed BezierAlign, which uses the control points of the curves to warp the word into a rectangular shape. TextDragon [7] predicts a series of quadrangles that follow the shape of the words and use their own RoISlide to rectify the text.

A recent trend in the community is to use transformer-based [47] networks. The fully connected topology of the self-attention mechanism avoids having to use any RoI rectifying operations at all. Some of them are capable of using simple annotations such as central keypoints [18,38]. For example, TTS [19] uses a shared transformer encoder-decoder and different decoder heads to perform word recognition, detection and segmentation. This method can be trained by either providing the ground truth polygon annotations, the bounding box, or only the text in the image. TESTR [52] also employs an encoder-decoder approach to perform text detection and recognition. The authors use two transformer decoder networks to extract the detection and the recognition. More recently, DEER [18] uses a transformer encoder to perform detection-agnostic detection and recognition. SwinTextSpotter [13] uses diverse transformer-encoder networks to improve the synergy between the detected areas and the recognizer.

2.2 Transformer Acceleration

The $O(n^2)$ complexity of the transformer architecture proposed by Vaswani et al. [47] has motivated multiple efforts to reduce its time and space complexity. On the NLP domain, different approaches have exploited the sparsity of the attention mechanism to reduce its complexity. For example, the Star-Transformer [10] replaces the fully connected attention with a star-shaped topology, reducing the complexity from quadratic to linear. Sparse Transformers [14] introduce multiple novel architectures that use sparse attention layers that perform faster un-batched decoding. The authors of the Linformer [50] also achieve linear complexity by approaching the self-attention with a low-rank matrix. Other networks such as TinyBERT [15] use distillation to transfer knowledge from a larger teacher BERT [4] network into a smaller one.

On the vision domain, numerous works have approached the problem by reducing the number of tokens on the input of the standard Vision Transformer (ViT) [5] architecture. The Hierarchical Vision Transformer [36] proposes an architecture that fuses tokens using a pooling operation after every transformer block, similar to the down-sampling of a convolutional network. EViT [22] progressively reduces the number of tokens along the different attention layers. The model uses the attention over the classification token to fuse uninformative tokens. DynamicViT [41] proposes a prediction module that estimates the importance of each token and discards tokens that are uninformative. Similar to EViT, the authors of ATS [6] propose a an adaptive token sampling method that uses the attention over the classification token to discard tokens. Unlike EViT, the method proposed is plug-and-play and does not need to be retrained. IA-RED^2 [35] employs a similar strategy to EViT, but they use a reinforcement algorithm to train the pruning algorithm.

3 Methodology

Transformer acceleration methods for vision are mostly focused on object classification with ViT-based models. By contrast, our method has been specifically designed with scene text understanding in mind. We test our proposed acceleration approach on our own transformer encoder-based model, which has been designed for fast inference speeds. The architecture performs detection and recognition using a shared convolutional backbone and two separate branches for detection and recognition. The detection branch is based on the Center-Net [53] architecture, which we use to predict the location of the text. These predicted locations are used to guide the pruning of uninformative image features before feeding them to the recognition branch, which uses the transformer encoder. This branch is capable of reading text in multiple orientations as well as curved text without performing RoI cropping or corrective operations. The self-attention mechanism of the transformer combines local information and encodes latent representations of the words in a grid (Fig. 5 shows an example of the recognition grid).

Our token pruning mechanism allows faster training and inference speeds without compromising the accuracy of the network, in addition of using less memory. By making use of large, publicly available datasets for scene-text detection and recognition, our approach is able to balance competitive quantitative results with fast inference speeds. Figure 3 shows a more detailed overview of the proposed architecture.

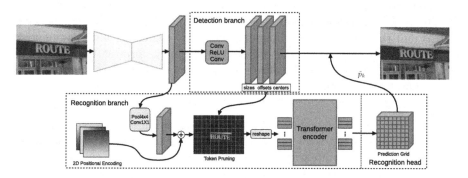

Fig. 3. A more detailed overview of the detection and recognition branches of our proposed architecture. The CenterNet-based detection branch generates detection proposals while the transformer-based recognition branch encodes a latent representation of each word in a grid. A separate recognition head generates a dense prediction map. The pruning mechanism uses the proposals from the detection branch to reduce the number of tokens at the input of the transformer encoder.

3.1 Architecture

Our model uses a ResNet-34 and a series of transposed convolutional operators as the backbone of the network. More specifically, we apply two transposed convolutions that expand the feature maps up to 1/8 of the original resolution. Inspired by U-Net [43], we combine the lower-level feature maps of the ResNet with up-scaled feature maps of the expanding path. In the original U-Net, the higher-level feature maps are cropped and then concatenated to the feature maps of the expansive path. Instead, our architecture applies a 1×1 convolutional operator over the ResNet feature maps to match the number of channels of the corresponding up-sampled maps, which are then added. The backbone outputs a feature map $f = \mathbb{R}^{\frac{W}{S} \times \frac{H}{S} \times D}$, where H and W are the height and width of the original image and S is the stride, in our case $S = 8$. The output feature map is shared by two different branches that perform detection and recognition.

Text Detection Branch. The text detection branch is based on the CenterNet architecture, an efficient object detection framework that predicts axis-aligned bounding boxes. CenterNet uses central keypoint estimation to predict the center of the bounding boxes by generating in a heatmap $\hat{Y} \in [0,1]^{\frac{W}{S} \times \frac{H}{S}}$. During training, the ground truth heatmap Y is generated by drawing a Gaussian kernel at the center of each object, which reduces the penalty around the ground truth keypoints. In this heatmap, a value $Y_{X,Y} = 1$ represents a keypoint, while $Y_{X,Y} = 0$ is background. The loss for the heatmap L_k is the modified focal loss [26] introduced by [20]. Following [20,53], we set $\alpha = 2$ and $\beta = 4$.

The stride of the feature map introduces a discretization error in the keypoint estimation. To overcome this, CenterNet introduces a local offset $\hat{O} \in \mathbb{R}^{\frac{W}{S} \times \frac{H}{S} \times 2}$ that helps to adjust each center. Like in the original CenterNet paper, the loss of the offsets L_{off} is the L1 loss at the keypoint locations. CenterNet predicts the widths and the heights of the bounding boxes by regressing both components at the center of each point, the output has the same form as the local offset $\hat{R} \in \mathbb{R}^{\frac{W}{S} \times \frac{H}{S} \times 2}$. The loss of the offsets L_{size} is again the L1 loss in each one of the ground truth keypoints.

Token Pruning. Our architecture introduces a token pruning strategy that reduce the number of tokens before the recognition transformer encoder. Since the space and time complexity of the attention mechanism of the transformer is quadratic with the number of input tokens, reducing the size of the input can yield more efficient training and inference times. This approach to reduce the number of tokens is based on the assumption that the features relevant to recognize text are local for each text instance, while the surrounding areas are uninformative. This strategy employs the detected text areas from the detection branch to discard part the visual features that come from the CNN. Any visual features from z_1 that does not overlap with a bounding box are discarded, since they probably do not contain textual information (Fig. 4 shows how features that are not overlapping any text detections get discarded). The pruning is applied after adding a 2D positional encoding [1,37] because we need to preserve the relative position of the tokens on the original feature map. Our experimental results show how the pruning mechanism does not affect the performance of the network, since all the relevant information for the recognition head is being preserved. This process is also fully end-to-end trainable.

This joint pruning and recognition strategy can be seen akin to two-stage architectures such as FOTS [27] or Mask TextSpotter [23,32], where each text detection is used to crop and recognize the RoI in the feature map. Our approach differs in that we do not crop and recognize the RoIs one by one. Instead, we remove the non-informative areas of the image features and perform the recognition in parallel. For a more fair comparison between both approaches, we also implemented a classic Two-Stage version of our model that performs RoI cropping and recognition with a transformer encoder. In our experimental section we show how the pruning approach yields faster training and inference times w.r.t. the Two-Stage version.

CNN features   Detector proposals   Pruned transformer input

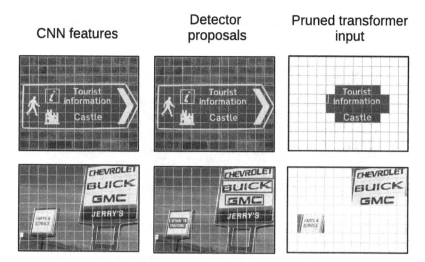

Fig. 4. Our pruning strategy discards tokens of the image that do not contain text information. Using the bounding boxes detected as a guide, we create a mask that has the size of the output feature map (which we abstractly represent as a grid over the original image), and use it to discard features that are outside of text regions. If the detector fails to localize a text instance, it will not get recognized.

Text Recognition Branch. The recognition head is principally composed of a transformer encoder [47]. In our experiments we have trained two different versions of the network, a Small version with 4 encoder layers and a Base version with 8 layers.

The inputs to the transformer are the up-scaled features outputted by the convolutional backbone in the form of a flattened vector of tokens. After applying the token pruning described in the previous section, the visual features are inputted into the transformer encoder. The self-attention mechanism of the transformer flexibly combines information around each cell of the input features to generate a latent representation grid.

After applying the self-attention, the recognition head generates a text prediction for every token, where the output is encoded as a maximum of M characters. We apply a softmax activation function over the character dimension to generate per-character confidences, obtaining final predictions of size $M \times C$, where C is the size of our alphabet. The loss of the recognition branch L_{recog} is the cross-entropy loss between the predicted character confidences and the ground truth one-hot vector of each character. Like the offset regression $\hat{O}$ and bounding box height and width regression, the loss for each word is calculated at the center of the words, while the predictions around it do not contribute to the loss.

The final optimization objective of the network is defined by the addition of the four previous losses:

$$L = L_k + L_{off} + \lambda_{size} L_{size} + L_{recog} \tag{1}$$

where λ_{size} is used to scale the bounding box regression loss, we set $\lambda_{size} = 0.1$ like in the original CenterNet paper.

Fig. 5. Visualization of the predicted word grid generated by the recognition head. The overlaid mask shows the confidence for each one of the predicted words (in this example, the prediction in blue represents the final recognized word). During training, the loss is only taken into account in the center of the word, while the areas around it are ignored. (Color figure online)

3.2   Training Details

The model was trained using two NVIDIA A40, the resolution during training was 1024×1024 with a batch size of 32. The optimizer used in all the cases was AdamW [31], with a gradual learning-rate warm-up of 1000 iterations. The model is pre-trained with SynthText for two epochs, with an initial learning rate of $1e-4$ with no learning rate decay. Next, the model is trained using a combined dataset of ICDAR13 [17], ICDAR15 [16], ICDAR17 [34], COCO-text [48] and TextOCR [46] for 40 epochs at the same learning rate. After 20 epochs we decay the learning rate to $1e-5$. Some datasets, such as ICDAR17, include text in different scripts than Latin, we do not take into account these text instances. During training, we use the ground truth detections to guide the token pruning.

4   Experiments

4.1   Text Detection and Recognition Datasets

We have evaluated our model on ICDAR13, ICDAR15, and Total-Text. On ICDAR13 we use the standard evaluation protocol proposed by the authors. The datasets ICDAR15 and Total-Text feature rotated quadrilaterals and irregular polygonal annotations respectively. To be able to compare our method in these datasets, we adopt the evaluation protocol proposed by TTS [19], where they propose to use the horizontal bounding box version of the ground truth annotation to evaluate the predictions. Their experimental results show that this evaluation strategy has a minor effect on the final results.

In the ICDAR15 and Total-Text datasets we used a single input resolution of 1400 and 2000 respectively. For ICDAR13 we use two different scales of 2000 and 500 pixels to better deal with text at different sizes.

Results. Table 1 shows end-to-end detection and recognition in quantitative results on ICDAR15. On this dataset our model manages to perform on par with the latest state-of-art models, which shows how the recognition branch is able to successfully deal with oriented text. Our model is also able of significantly higher inference speeds than the latest models and it is only surpassed by FOTS, which our model manages to widely surpass on accuracy.

Table 1. End-To-End results on ICDAR15. The results reported were obtained using the Small version of our network.

Method	IC15			FPS
	S	W	G	
FOTS [27]	81.1	75.9	60.8	**22.6**
Boundary [49]	79.7	75.2	64.1	–
TextPerceptron [40]	80.5	76.6	65.1	8.8
ABCnet [30]	82.7	78.5	73.0	10.0
TextDragon [7]	**86.2**	**82.0**	68.1	–
MANGO [39]	85.4	80.1	73.9	4.3
DEER [18]	82.7	79.0	75.6	–
SwinTextSpotter [13]	83.9	77.3	70.5	1.2
TESTR [52]	85.2	79.4	73.6	2.0
TTS [19]	85.0	81.5	**77.3**	–
Ours	84.6	80.2	71.6	18.0

In Table 2 we show the results for the datasets ICDAR13 and Total-Text. Our model obtains good results using two scales, achieving competitive results with the latest models. When using a single scale the performance drops, but still maintains good results. Despite never seeing the training set of Total-Text (a dataset mainly focused on rotated text), our model also obtains good results on it.

Finally, Fig. 6 shows qualitative examples of word detection and recognition with different types of distortions. Our model is capable of dealing with different types of distortions such as rotations or curvatures thanks to the transformer encoder-based recognizer.

Table 2. End-To-End results on ICDAR13 and Total-Text.

Method	IC13			Total-Text	
	S	W	G	None	full
FOTS [27]	88.8	87.1	86.0	–	–
Boundary [49]	88.2	87.7	84.1	65.0	–
TextPerceptron [40]	91.4	90.7	85.8	69.7	78.3
ABCnet [30]	–	–	–	70.4	78.1
TextDragon [7]	–	–	–	75.8	84.4
MANGO [39]	**93.4**	**92.3**	**88.7**	72.9	83.6
DEER [18]	–	–	–	74.8	81.3
SwinTextSpotter [13]	–	–	–	74.3	84.1
TESTR [52]	–	–	–	73.3	83.9
TTS$_{poly}$ [19]	–	–	–	75.6	84.4
TTS$_{box}$ [19]	–	–	–	**75.9**	**84.5**
Ours	85.2	83.4	78.3	61.5	72.1
Ours TS	92.3	89.2	87.2	64.2	74.6

(a) Examples of successful detection and recognition with the Small model.

Fig. 6. Our transformer-based approach manages to successfully perform detection and recognition with horizontal bounding boxes. The models successfully recognizes the text with different types of distortions.

4.2 Token Pruning

In this section we evaluate the performance gains of our token pruning approach. We also compare it with an alternative Two-Stage variation of our model. In this version, the locations of the detection branch are used to crop the image features overlapping the bounding boxes. The transformer encoder performs recognition for each one of the cropped regions, but unlike our architecture this does not happen in parallel. The three versions of our architecture were trained using the Small (with 4 attention layers) and Base (with 8 layers) sizes of the transformer encoder.

In Table 3 we compare the three variants of our architecture trained under the same configuration. As seen in the table, the three variants offer similar quantitative results on ICDAR15. The token pruning and Two-Stage variants considerably reduce the number of MAC operations (one multiplication and one addition) with respect the model with no pruning. At a resolution of 2000 pixels, the pruning mechanism removes an average of 91% of non-informative tokens, which reduces 95% of the operations in the transformer encoder. The number of operations remains similar between the Small and Base versions, which allows our pruning approach to maintain almost the same number of FPS. In the Base version of our model, the reduction represents almost half of the overall operations of the network (from 510 to 279 GMACs).

Since the recognition happens in parallel in the fully connected attention mechanism of the transformer encoder, the proposed pruning version is slightly more computationally expensive than the Two-Stage version. However, the parallelization of the transformer operations allows the pruning version to obtain faster inference speeds. The Base version of our model has almost the same number of operations as the Small one, and reaches similar FPS during inference.

Training. The benefit of this parallel approach is that is easier to batch the recognition during training, which results in reduced training times. In the rightmost column of Table 3 we can see the number of iterations per second for all the variants of our model using the same configuration (an input resolution of 1024×1024 and a batch size of 32). Using token pruning reduces 40% the training times for the Small model while in the Base model the reduction is 76%.

Image Resolution. A bigger image size has a quadratic impact on the number of operations of the encoder head. In Fig. 7 we can see the effect of input image size on the FPS of the three variations during inference. Thanks to the great reduction in number of operations of the pruning, our proposed strategy achieves faster inference speeds than the two variations of our model. The gap between the Small and Base non-pruning version (red lines) is considerably bigger than the gap between the two versions of our Pruning approach, which is between 1 and 2 FPS depending on the size.

Table 3. Performance comparison between using token pruning, no pruning and Two-Stage for inference. The models were evaluated on ICDAR13 at a resolution of 2000 pixels. The MACs count includes the operations of the convolutional backbone (which totals 276 GMACs for the used image size). The rightmost column also shows the number of iterations per second during training. The batch size is 32 and the resolution used is 1024 × 1024.

Recognition	Layers	S	W	G	FPS	GMACs	it/s
No Pruning	4	79.3	77.1	74.1	12	394	0.84
Two-Stage	4	**79.5**	**77.2**	74.5	15	**278**	1.06
Pruning	4	79.4	77.1	**74.7**	**18**	282	**1.18**
No Pruning	8	79.1	77.2	74.2	8	510	0.65
Two-Stage	8	79.1	**77.4**	74.4	12	**279**	1.02
Pruning	8	**79.3**	77.3	**75.1**	**17**	284	**1.15**

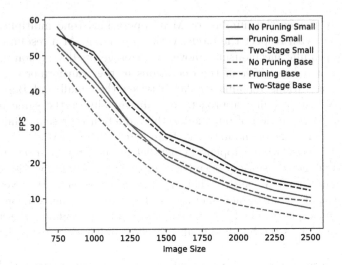

Fig. 7. Effects of the input image size on the FPS for the different variants of our model. The solid line shows the performance using the Small transformer (4 encoder layers) while the dashed line shows the performance of the Base model (8 encoder layers). (Color figure online)

5   Conclusions

In this paper we have introduced a novel strategy to improve the efficiency of transformer-based architectures for scene text recognition. Our token pruning mechanism, which has been specially designed for scene-text detection and recognition, effectively decreases training and inference times of the network. We have tested this approach on our own transformer-based architecture, which has been tailored to achieve a balance between speed and accuracy. Thanks to the proposed pruning mechanism, our model achieves fast inference speeds while being competitive with the state of the art.

Acknowledgements. This work has been supported by grants PDC2021-121512-I00, PID2020-116298GB-I00 and PLEC2021-007850 funded by the European Union NextGenerationEU/PRTR and MCIN/AEI/10.13039/501100011033; the EU Lighthouse on Safe and Secure AI - ELSA funded by European Union's Horizon Europe programme under grant agreement No 101070617; the Spanish Project NEOTEC SNEO-20211172 from CDTI; grant Torres Quevedo PTQ2019-010662; and the Industrial Doctorate programme of the Catalan Government (2020 DI 058).

References

1. Bello, I., Zoph, B., Vaswani, A., Shlens, J., Le, Q.V.: Attention augmented convolutional networks. In: Proceedings of the IEEE/CVF International Conference on Computer Vision, pp. 3286–3295 (2019)

2. Biten, A.F., et al.: Scene text visual question answering. In: Proceedings of the IEEE/CVF International Conference on Computer Vision, pp. 4291–4301 (2019)
3. Ch'ng, C.K., Chan, C.S.: Total-text: a comprehensive dataset for scene text detection and recognition. In: 2017 14th IAPR International Conference on Document Analysis and Recognition (ICDAR), vol. 1, pp. 935–942. IEEE (2017)
4. Devlin, J., Chang, M.-W., Lee, K., Toutanova, K.: Bert: pre-training of deep bidirectional transformers for language understanding. arXiv preprint arXiv:1810.04805 (2018)
5. Dosovitskiy, A., et al.: An image is worth 16x16 words: transformers for image recognition at scale. arXiv preprint arXiv:2010.11929 (2020)
6. Fayyaz, M., et al.: Adaptive token sampling for efficient vision transformers. In: Avidan, S., Brostow, G., Cissé, M., Farinella, G.M., Hassner, T. (eds.) Computer Vision - ECCV 2022. ECCV 2022. Lecture Notes in Computer Science, vol. 13671, pp. 396–414. Springer, Cham (2022). https://doi.org/10.1007/978-3-031-20083-0_24
7. Feng, W., He, W., Yin, F., Zhang, X.Y., Liu, C.-L.: Textdragon: an end-to-end framework for arbitrary shaped text spotting. In: Proceedings of the IEEE/CVF International Conference on Computer Vision, pp. 9076–9085 (2019)
8. Gómez, L., Rusinol, M., Karatzas, D.: Cutting sayre's knot: reading scene text without segmentation. application to utility meters. In: 2018 13th IAPR International Workshop on Document Analysis Systems (DAS), pp. 97–102. IEEE (2018)
9. Graves, A., Fernández, S., Gomez, F., Schmidhuber, J.: Connectionist temporal classification: labelling unsegmented sequence data with recurrent neural networks. In: Proceedings of the 23rd International Conference on Machine Learning, pp. 369–376 (2006)
10. Guo, Q., Qiu, X., Liu, P., Shao, Y., Xue, X., Zhang, Z.: Star-transformer. arXiv preprint arXiv:1902.09113 (2019)
11. He, T., Tian, Z., Huang, W., Shen, C., Qiao, Y., Sun, C.: An end-to-end textspotter with explicit alignment and attention. In: Proceedings of the IEEE Conference on Computer Vision and Pattern Recognition, pp. 5020–5029 (2018)
12. Hochreiter, S., Schmidhuber, J.: Long short-term memory. Neural Comput. **9**(8), 1735–1780 (1997)
13. Huang, M., et al.: Swintextspotter: scene text spotting via better synergy between text detection and text recognition. In: Proceedings of the IEEE/CVF Conference on Computer Vision and Pattern Recognition, pp. 4593–4603 (2022)
14. Jaszczur, S., et al.: Sparse is enough in scaling transformers. Adv. Neural Inf. Process. Syst. **34**, 9895–9907 (2021)
15. Jiao, X., et al.: Tinybert: distilling bert for natural language understanding. arXiv preprint arXiv:1909.10351 (2019)
16. Karatzas, D., et al.: ICDAR 2015 competition on robust reading. In: 2015 13th International Conference on Document Analysis and Recognition (ICDAR), pp. 1156–1160. IEEE (2015)
17. Karatzas, D., et al.: ICDAR 2013 robust reading competition. In: 2013 12th International Conference on Document Analysis and Recognition, pp. 1484–1493. IEEE (2013)
18. Kim, S., et al.: Deer: detection-agnostic end-to-end recognizer for scene text spotting. arXiv preprint arXiv:2203.05122 (2022)
19. Kittenplon, Y., Lavi, I., Fogel, S., Bar, Y., Manmatha, R., Perona, P.: Towards weakly-supervised text spotting using a multi-task transformer. arXiv preprint arXiv:2202.05508 (2022)

20. Law, H., Deng, J.: Cornernet: detecting objects as paired keypoints. In: Proceedings of the European Conference on Computer Vision (ECCV), pp. 734–750 (2018)
21. Lee, C.-Y., Osindero, S.: Recursive recurrent nets with attention modeling for OCR in the wild. In: Proceedings of the IEEE Conference on Computer Vision and Pattern Recognition, pp. 2231–2239 (2016)
22. Liang, Y., Ge, C., Tong, Z., Song, Y., Wang, J., Xie, P.: Not all patches are what you need: expediting vision transformers via token reorganizations. arXiv preprint arXiv:2202.07800 (2022)
23. Liao, M., Pang, G., Huang, J., Hassner, T., Bai, X.: Mask TextSpotter v3: segmentation proposal network for robust scene text spotting. In: Vedaldi, A., Bischof, H., Brox, T., Frahm, J.-M. (eds.) ECCV 2020. LNCS, vol. 12356, pp. 706–722. Springer, Cham (2020). https://doi.org/10.1007/978-3-030-58621-8_41
24. Liao, M., Shi, B., Bai, X.: Textboxes++: a single-shot oriented scene text detector. IEEE Trans. Image Process. 27(8), 3676–3690 (2018)
25. Liao, M., Shi, B., Bai, X., Wang, X., Liu, W.: Textboxes: a fast text detector with a single deep neural network. In: Thirty-First AAAI Conference on Artificial Intelligence (2017)
26. Lin, T.-Y., Goyal, P., Girshick, R., He, K., Dollár, P.: Focal loss for dense object detection. In: Proceedings of the IEEE International Conference on Computer Vision, pp. 2980–2988 (2017)
27. Liu, X., Liang, D., Yan, S., Chen, D., Qiao, Y., Yan, J.: Fots: fast oriented text spotting with a unified network. In: Proceedings of the IEEE Conference on Computer Vision and Pattern Recognition, pp. 5676–5685 (2018)
28. Liu, Y., Chen, H., Shen, C., He, T., Jin, L., Wang, L.: Abcnet: real-time scene text spotting with adaptive bezier-curve network. In: Proceedings of the IEEE/CVF Conference on Computer Vision and Pattern Recognition, pp. 9809–9818 (2020)
29. Liu, Y., Jin, L., Zhang, S., Luo, C., Zhang, S.: Curved scene text detection via transverse and longitudinal sequence connection. Pattern Recogn. 90, 337–345 (2019)
30. Liu, Y., et al.: Abcnet v2: adaptive bezier-curve network for real-time end-to-end text spotting. arXiv preprint arXiv:2105.03620 (2021)
31. Loshchilov, I., Hutter, F.: Decoupled weight decay regularization. arXiv preprint arXiv:1711.05101 (2017)
32. Lyu, P., Liao, M., Yao, C., Wu, W., Bai, X.: Mask textspotter: an end-to-end trainable neural network for spotting text with arbitrary shapes. In: Proceedings of the European Conference on Computer Vision (ECCV), pp. 67–83 (2018)
33. Mafla, A., Dey, S., Biten, A.F., Gomez, L., Karatzas, D.: Fine-grained image classification and retrieval by combining visual and locally pooled textual features. In: Proceedings of the IEEE/CVF Winter Conference on Applications of Computer Vision, pp. 2950–2959 (2020)
34. Nayef, N., et al.: ICDAR 2017 robust reading challenge on multi-lingual scene text detection and script identification-RRC-MLT. In: 2017 14th IAPR International Conference on Document Analysis and Recognition (ICDAR), vol. 1, pp. 1454–1459. IEEE (2017)
35. Pan, B., et al.: IA-RED^2: interpretability-aware redundancy reduction for vision transformers. Adv. Neural Inf. Process. Syst. 34, 24898–24911 (2021)
36. Pan, Z., Zhuang, B., Liu, J., He, H., Cai, J.: Scalable vision transformers with hierarchical pooling. In: Proceedings of the IEEE/CVF International Conference on Computer Vision, pp. 377–386 (2021)
37. Parmar, N., et al.: Image transformer. In: International Conference on Machine Learning, pp. 4055–4064. PMLR (2018)

38. Peng, D., et al. Spts: single-point text spotting. arXiv preprint arXiv:2112.07917 (2021)
39. Qiao, L., et al.: Mango: a mask attention guided one-stage scene text spotter. arXiv preprint arXiv:2012.04350 (2020)
40. Qiao, L., et al.: Text perceptron: towards end-to-end arbitrary-shaped text spotting. In: Proceedings of the AAAI Conference on Artificial Intelligence, vol. 34, pp. 11899–11907 (2020)
41. Rao, Y., Zhao, W., Liu, B., Jiwen, L., Zhou, J., Hsieh, C.-J.: Dynamicvit: efficient vision transformers with dynamic token sparsification. Adv. Neural Inf. Process. Syst. **34**, 13937–13949 (2021)
42. Reddy, S., Mathew, M., Gomez, L., Rusinol, M., Karatzas, D., Jawahar, C.V.: Roadtext-1k: text detection & recognition dataset for driving videos. In: 2020 IEEE International Conference on Robotics and Automation (ICRA), pp. 11074–11080. IEEE (2020)
43. Ronneberger, O., Fischer, P., Brox, T.: U-net: convolutional networks for biomedical image segmentation. In: Navab, N., Hornegger, J., Wells, W.M., Frangi, A.F. (eds.) MICCAI 2015. LNCS, vol. 9351, pp. 234–241. Springer, Cham (2015). https://doi.org/10.1007/978-3-319-24574-4_28
44. Schuster, M., Paliwal, K.K.: Bidirectional recurrent neural networks. IEEE Trans. Sig. Process. **45**(11), 2673–2681 (1997)
45. Singh, A., et al.: Towards VQA models that can read. In: Proceedings of the IEEE/CVF Conference on Computer Vision and Pattern Recognition, pp. 8317–8326 (2019)
46. Singh, A., Pang, G., Toh, M., Huang, J., Galuba, W., Hassner, T.: Textocr: towards large-scale end-to-end reasoning for arbitrary-shaped scene text. In: Proceedings of the IEEE/CVF Conference on Computer Vision and Pattern Recognition, pp. 8802–8812 (2021)
47. Vaswani, A., et al: Attention is all you need. Adv. Neural Inf. Process. Syst. **30** (2017)
48. Veit, A., Matera, T., Neumann, L., Matas, J., Belongie, S.: Coco-text: dataset and benchmark for text detection and recognition in natural images. arXiv preprint arXiv:1601.07140 (2016)
49. Wang, H., et al.: All you need is boundary: toward arbitrary-shaped text spotting. In: Proceedings of the AAAI Conference on Artificial Intelligence, vol. 34, pp. 12160–12167 (2020)
50. Wang, S., Li, B.Z., Khabsa, M., Fang, H., Ma, H.: Linformer: self-attention with linear complexity. arXiv preprint arXiv:2006.04768 (2020)
51. Xing, L., Tian, Z., Huang, W., Scott, M.R.: Convolutional character networks. In: Proceedings of the IEEE/CVF International Conference on Computer Vision, pp. 9126–9136 (2019)
52. Zhang, X., Su, Y., Tripathi, S., Tu, Z.: Text spotting transformers. In: Proceedings of the IEEE/CVF Conference on Computer Vision and Pattern Recognition, pp. 9519–9528 (2022)
53. Zhou, X., Wang, D., Krähenbühl, P.: Objects as points. arXiv preprint arXiv:1904.07850 (2019)

Text Enhancement: Scene Text Recognition in Hazy Weather

En Deng, Gang Zhou[✉], Jiakun Tian, Yangxin Liu, and Zhenhong Jia

Key Laboratory of Signal Detection and Processing, School of Information Science and Engineering, Xinjiang University, Xinjiang, People's Republic of China
`gangzhou_xju@126.com, jzhh@xju.edu.cn`

Abstract. Although deep learning-based methods for scene text recognition have achieved remarkable results on conventional datasets, recognizing text images under adverse weather conditions with poor visibility remains challenging. To address this problem, we propose a text image enhancement network that can be embedded into a scene text recognizer in a pluggable manner. This network comprises multiple sets of digital image processing (DIP) units, which are composed of differentiable filters whose parameters are estimated by a parallel sequential residual block (PSRB). The PSRB is a novelty designed module for extracting the sequence information of text image in a parallel manner. The whole network can be trained end-to-end with a text-image enhancement module and a recognition module. Extensive experiments on synthetic and real-world benchmarks demonstrate that our approach significantly improves the performance of popular text recognizers in adverse weather conditions.

Keywords: Scene Text Recognition · Image Enhancement · Parallelly Sequence Residual Block

1 Introduction

Scene text recognition has been extensively applied in various fields, such as license plate recognition [1]and document retrieval [2], highlighting its significance in the field of computer vision. However, the adverse weather, such as haze, has a negative influence on outdoor scene text recognition. It is attributed to image degration in haze weather, and weakens the effectiveness of many existing scene text recognition methods.

Some text recognition methods [3–6] have achieved impressive results on clear images, and have exceeded 90% recognition rates on standard datasets. However, the performance of scene text recognition methods drastically drops in adverse

Supported by the National Natural Science Foundation of China under grant No. 62166040, 62261053, 62137002, Natural Science Foundation of Xinjiang Autonomous Region under grant No. 2021D01C057, and the National Key R&D Program of China under grant No. 2021ZD0113601.

G. A. Fink et al. (Eds.): ICDAR 2023, LNCS 14192, pp. 122–136, 2023.
https://doi.org/10.1007/978-3-031-41731-3_8

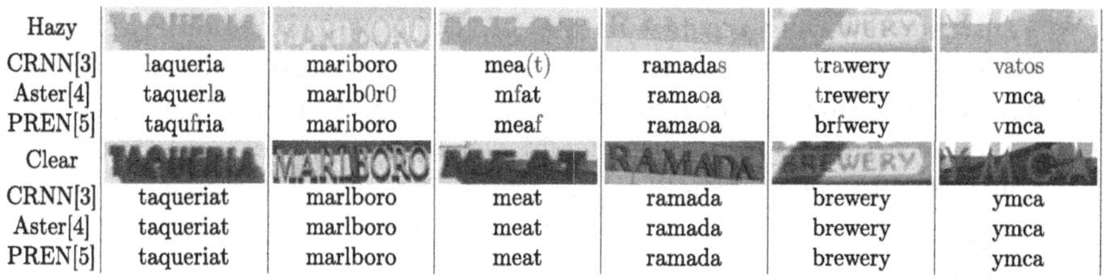

Hazy						
CRNN[3]	laqueria	mariboro	mea(t)	ramadas	trawery	vatos
Aster[4]	taquerla	marlb0r0	mfat	ramaoa	trewery	vmca
PREN[5]	taqufria	mariboro	meaf	ramaoa	brfwery	vmca
Clear						
CRNN[3]	taqueriat	marlboro	meat	ramada	brewery	ymca
Aster[4]	taqueriat	marlboro	meat	ramada	brewery	ymca
PREN[5]	taqueriat	marlboro	meat	ramada	brewery	ymca

Fig. 1. Recognition comparison between different recognizers in haze images or clear images. Those characters in red denote wrong recognition. (Color figure online)

weather conditions. As shown in Fig. 1, existing scene text recognition methods are not applicable under adverse weather conditions. Adverse weather conditions can cause degradation of text images, which in turn impacts the extraction of text features, resulting in the failure of existing scene text recognition methods.

Recent image dehazing networks [7–9] have achieved remarkable success, these methods target high resolution images and aim at pixel-level recovery. However, the cropped scene text images lack prior information, making it challenging to perform image dehazing. Moreover, text recognition is a high-level vision task that is hard to improve by pixel-level recovery. Therefore these dehazing methods are not applicable to scene text recognition. To address the challenge of low resolution and text blurring in scene text recognition, recent advancements in text super-resolution methods [10–13] have emerged as a potential preprocessing solution. Inspired by these text super-resolution methods, we believe that image enhancement can be used as a pre-processing method to improve text recognition in hazy weather scenes.

Compared to general image enhancement methods [14,15], text image enhancement requires consideration of sequential relevance among text. Therefore, we propose a dedicated text image enhancement network, followed by utilizing a pre-trained text recognition model for calculating recognition loss. As shown in Fig. 2. Our network comprises three main components: DIP units, PSRB, and a pre-trained recognition network. The PSRB is for predicting the parameters of the differentiable filter. It is able to capture textual information and the character stroke detail information. As a result, the multiple DIP units enhance text images for recognition. To the best of our knowledge, we are the first to discuss that the use of image enhancement networks for improving the recognition accuracy of text images in hazy weather.

The contributions of this work are therefore three-fold:

- We propose a novel text enhancement network for addressing scene text recognition in adverse weather conditions. This network comprises multiple DIP units and a scene text recognition module.
- To achieve optimal parameter guidance for differentiable filters, our proposed PSRB is able to efficiently capture the horizontal and vertical correlations of characters in text images.

- We constructed a synthetic haze dataset of 27k images from standard datasets. Additionally, we collected and annotated a set of real-to-world haze images. Extensive experiments were conducted on this dataset to evaluate the performance of our proposed method and other existing methods. The results demonstrated the effectiveness of our proposed text image enhancement approach.

2   Related Work

2.1   Scene Text Recognition

The task of text recognition is to identify the corresponding sequence of strings from the cropped blocks of text images. According to a recent literature review [16], early scene text recognition algorithms were mainly based on segmentation approaches. These algorithms located each character from a text image, and recognized characters one by one. With the development of deep learning, scene text recognition have made significant progress in recent years. The CRNN [3,17,18] framework encodes images as hidden vectors, and decodes them by concatenated temporal classification (CTC) [19]. This approach eliminates the need for tedious character-by-character segmentation and effectively improves the performance of natural scene text recognition algorithms through CTC decoding. Recently, Attention-based frameworks have also shown promising results in recent recognition methods. ASTER [4] preprocesses a recognizer with STN [20] networks to correct curved text into horizontal text, and utilizes a bidirectional attention decoder to achieve better robustness on irregular scene text. However, attention-based frameworks often suffer from attention drift, which means that the attention model cannot accurately associate each feature vector with the corresponding target region in the input image. Li et al. [21] attempts to directly focus on the spatial relationships of 2-dimensional feature maps by a 2-dimensional attention mechanism. Wang et al. [22] proposes to attention decoupling network for the attention drift problem by a convolutional alignment module that performs alignment with the visual features of the encoder. Lyu et al. [23] propose a relational attention module and a parallel attention module to obtain 2-dimensional spatial information of text. Lee et al. [24] introduces a self-attention mechanism to handle the edge cases in scene text recognition, and the decoder can input 2-dimensional feature maps. This allows for obtaining long-range dependencies through the self-attentive layer, enabling consecutive character localization and achieving better performance for multi-line text. Lu et al. [6] uses an encoder module based on multifaceted global contextual attention and a decoder module based on the transformer [25]. This model is capable of learning both input-output attention and self-attention, where self-attention is composed of encoding features and target relations within the encoder and decoder. Deep learning-based scene text recognition methods have achieved over 90% accuracy on clear scene text datasets.

However, scene text recognition is negatively influenced by image degradation, such as camera shaking and low-resolution ratio. Mou et al. [10] proposed a degradation-aware scene text recognizer with embedded super-resolution units for recognizing low-resolution scene text. Wang et al. [26] introduced a text super-resolution network that captures the sequence information of text images by adding two recursive layers in the backbone. Chen et al. [11] presents the text prior to the model by proposing a position-aware module and a content-aware module. Ma et al. [12] merges text-specific semantic features into each block in the backbone and uses an iterative approach to enhance text images. Chen et al. [27] proposed a super-resolution method for text images of stroke-aware scenes based on Gestalt psychology. This method designs rules for identifying English letters and numbers at the stroke level, which provides finer-grained attentional guidance highlighting the stroke region's details. Although the super-resolution network solves the problem of low-resolution scene text recognition to a certain extent, the degradation of scene text images caused by adverse weather is different. Because of atmospheric light scattering interference the appearance and contrast of the text in the image is reduced.

2.2   Image Dehazing

Recently, various approaches [28–30] have been proposed to directly recover haze images. For example, Wu et al. [31] proposed a novel contrast regularization method based on contrastive learning, which ensures that the restored image is pulled closer to the original clear image. However, single image dehazing is aimed at restoring hazy images to haze-free images at the pixel-level as much as possible, without considering adaptability for advanced tasks. Recently, there are many researchers realized that image recovery does help the object detection in hazy weather [32,33]. Li et al. [34] proposed a light-weight dehazing model AOD-Net, which cascaded a object detection model into a unified network to achieve object detection under hazy conditions. Liu et al. [35] proposed an image adaptive target detection method, which designed a differentiable image processing module to achieve image enhancement and end-to-end training with a subsequent target detection model. In the hand-craft feature era, there was some literatures [36–38] discussing license plate recognition in extreme weather. In [39] normalized license plates with overcast weather information were used and then a Gaussian filter was used to enhance the image quality. Various image processing methods such as smoothing filters, edge detection, adaptive thresholding, and local feature extraction were used to improve the accuracy of license plate recognition. Inspired by the above methods, we adopt scene text recognition method based on image enhancement. The image enhancement module, such as image enhancement, and the pre-trained text recognition module are unified under a holistic framework to achieve a unified optimization goal for both vision tasks.

3   Proposed Method

In this section, we present our proposed text enhancement network, which aims to improve text image quality while enhancing recognition accuracy. The network architecture is shown in Fig. 2, and it consists of three main components: the PSRB, DIP units, and pre-trained recognizer. The PSRB is responsible for predicting the parameters of differentiable filters in the DIP units based on the input cropped text image. Each DIP unit independently generates an enhanced image, which is then summed to obtain the final output. To generate text-focused enhanced images, we utilize a pre-trained recognition network to compute recognition loss on the enhanced images.

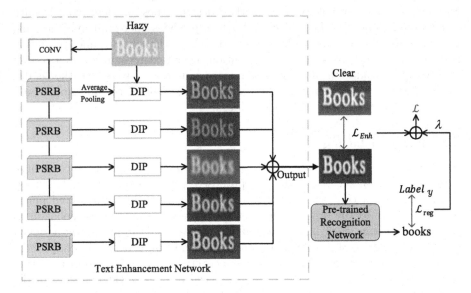

Fig. 2. The overall architecture of our text enhancement networks consists of three parts, including a PSRB, a DIP units, and a pre-trained recognition module. Parameters in the pre-trained recognition are frozen.

3.1   Overview of Text Enhancement Network

The overall architecture of the network is illustrated in Fig. 2, and it consists of multiple filter in parallel. Specifically, the top branch of the network employs a convolutional neural network (CNN) to extract coarse features that facilitates PSRB to extract more textual details. The PSRB estimates the parameters for each enhancement branch, which consists of a group of filters that output an enhanced image individually. By normalization the enhanced images from all branches, the final output can effectively removing haze and improving the visual quality of the text images. Moreover, the enhanced images are inputted into a pre-trained recognition network, where cross-entropy loss is employed to distinguish potentially confusable characters.

The objective of image enhancement is to improve the visual quality and legibility of images that are captured in haze conditions. The process typically involves training a set of filters that can remove the interference of environmental factors and highlight the important features for recognition. Inspired by IA-YOLO [35], a novelty image enhancement framework is proposed, which enables the backpropagation to update the parameters. This framework is composed of multilayer filters, including White Balance (WB), Gamma, Contrast, Tone, and Sharpen. Specifically, WB adjusts the color balance of the image, Gamma modifies the brightness, Contrast enhances the differences between the light and dark regions, Tone adjusts the overall tonality of the image, they can be expressed as pixel-level filters. Sharpen increases the text edge contrast. The mathematical expressions for these pixel-level filters are shown in Table 1.

Table 1. The functions of pixel-wise filters.

Recognizer	Parameters	Mapping Function
WB	W_r, W_g, W_b	$P_o = (W_r r_i, W_g g_i, W_b b_i)$
Gamma	G	$P_o = P_i^G$
Contrast	α	$P_o = \alpha \cdot E_n(P_i) + (1 - \alpha) \cdot P_i$
Tone	t_i	$P_o = \left(L_{t_r}(r_i), L_{t_g}(g_i), L_{t_b}(b_i)\right)$

The pixel-wise filters map the input pixel values to the output pixel values, representing the values red, green, and blue for each of the three color channels. Table 1 lists the mapping functions of the four pixel-wise filters, where the second column lists the parameters to be optimized in our approach. WB and Gamma are simple multiplication and power transformation. The differentiable contrast filter sets a linear interpolation between the original image and the fully enhanced image by the input parameters. As shown in Table 1, $E_n(P_i)$ in the mapping function is defined as follows:

$$Lum(P_i) = 0.27r_i + 0.67g_i + 0.06b_i \tag{1}$$

$$EnLum(P_i) = \frac{1}{2}(1 - \cos(\pi \times Lum(P_i))) \tag{2}$$

$$E_n(P_i) = P_i \times \frac{EnLum(P_i)}{Lum(P_i)} \tag{3}$$

The Tone filter is designed as a monotonic segmented linear function. Learn the Tone filter with L parameters, denoted as $\{t_0, t_1, \ldots, t_{L-1}\}$. The points of the tone curve are denoted as $(k/L, T_k/T_L)$ where $T_k = \sum_{i=0}^{k-1} t_l$. In addition, the expressed mapping function is differentiable in terms of parameters, which makes the differentiable function on the input image and parameters $\{t_0, t_1, \ldots, t_{L-1}\}$ as follows:

$$P_o = \frac{1}{T_L} \sum_{j=0}^{L-1} clip(L \cdot P_i - j, 0, 1) t_k \tag{4}$$

Image sharpening can highlight image text details and improve the overall visual quality of an image.

$$F(x, \lambda) = I(x) + \lambda(I(x) - Gau(I(x))) \tag{5}$$

where $I(x)$ is the input image, $Gau(I(x))$ is the Gaussian filter, and λ is the positive scale factor. Unlike IA-YOLO [35] we do not need to downsample the images because our images are all 32×128 in size.

3.2 Parallelly Sequential Residual Block (PSRB)

In order to predict the DIP parameters, the visual coder is required to sense the image's global information through the use of differentiability filters. However, text images have strong sequential characteristics that need to be considered. Textzoom [26] has sought to address this issue by utilizing contextual information in both horizontal and vertical directions to extract such relationships. However, their approach only models the two sets of context dependencies in a sequential manner. An improvement to this method can be made by utilizing recursive connections of visual features in both horizontal and vertical directions using an RNN. Such an approach would better capture contextual relationships within text images.

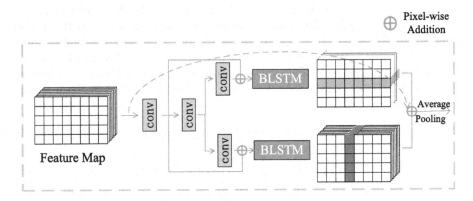

Fig. 3. Parallelly Sequential Residual Block (PSRB).

Based on Textzoom [26], we propose a bidirectional sequential residual parameter predictor to extract feature dependencies in both vertical and horizontal directions. As shown in Fig. 3, specifically, the haze image is convolved to obtain shallow features, followed by a CNN layer. We choose LSTM units as the basic units of RNN. The bidirectional RNNs are employed to construct sequence-related features in both the vertical and horizontal directions, while cyclically updating their internal states in the hidden layer.

$$H_{t1} = \varphi\left(H_{t1-1}, (X_r + X_h)\right), \quad t1 = 1, 2 \ldots, W \tag{6}$$

$$H_{t2} = \varphi\left(H_{t2-1}, (X_r + X_v)\right), \quad t2 = 1, 2 \dots, H \tag{7}$$

where H_t represents the RNN hidden unit, which X_r, X_h, X_v respectively denote the trunk, horizontal and vertical features processed by the CNN layers. Specifically, horizontal direction modeling is used to construct character-to-character dependencies. In contrast, vertical direction modeling is used for textual context within characters. We obtain two directional feature maps connected with the original feature maps for residuals to capture contextual information in haze text images more effectively.

$$F_{out} = O + O_k + O_v \tag{8}$$

where O denotes the shallow features of the haze image and O_k, O_v respectively denotes the recursively connected RNN output features in the horizontal and vertical directions. To obtain the final output feature map of the bi-directional sequential residual parameter predictor, we apply global average pooling to it, which results in an output of size $1 \times 1 \times 64$. Finally, we use fully connected layers to predict the parameters for the filters.

3.3   Overall Loss Function

In the task of enhancing text images, character regions are usually given more attention than the background. However, using simple L1 or L2 loss to train the network may only improve the detailed texture of the image, without considering the content of the text itself. This approach is effective for regular images, but not practical for text images where the content is more important than the texture. Wang et al. [40] proposed a perceptual loss, which utilizes a pre-trained VGG [41] network to calculate the similarity between the feature maps of the enhanced and original images. By pre-trained on ImageNet, which contains 1000 objects, the VGG network can help the network to understand the general content of the image. Additionally, we pre-trained a Transformer-based recognition model using synthetic text datasets (including Syn90k [42] and SynthText [43]). The recognition model is then used to supervise the text prediction using cross-entropy loss. Since the parameters of the pre-trained transformer are frozen, the recognition module will more attention to text images confusable characters. Therefore, the overall loss function is defined as:

$$\mathcal{L} = \mathcal{L}_{Enh} + \lambda_{rec}\mathcal{L}_{rec} \tag{9}$$

where λ_{rec} is the weight that balances these two loss functions.

4   Experiments

4.1   Dataset

In this paper, we train our enhancement model only on one synthetic datasets without any finetuning on other real datasets. We evaluate our model on three standard benchmarks that contain two synthetic haze scene text datasets and one real-to-word haze scene text datasets.

ICDAR2017MLT [44]: ICDAR2017MLT contains 7200 training images and 1800 test images. We use the atmospheric scattering model to randomly generate haze with different concentrations and then crop it according to the text labels. We ignore the images with less than three characters according to the general settings. A total of 27000 pairs of training text images and 6266 pairs of test images are obtained.

Street View Text (SVT) [45]: The SVT contains 350 images taken by Google Street View, and the use atmospheric scattering model generates haze randomly with different concentrations. 257 images for training and 647 images for testing can be obtained after cropping, and only the test images are selected for our experiments.

Real-to-world: 204 real haze images containing text are collected on the web, and 388 images can be obtained for testing after manual annotation and cropping.

4.2 Implementation Details

In the image processing stage, we follow the general image recognition processing method by unifying the size of all haze images and clear images to 32×128. During the training period, the learning rate is fixed to $1e-4$, and through ablation studies, we set the trade-off weight of λ_{rec} as 0.5. We use the Adam optimizer with a momentum term of 0.9. The parameters in the pre-trained recognizer are frozen. In evaluating recognition accuracy, we use the officially released ASTER[1] [4], CRNN[2] [3], and MORAN[3] [46] models. All the enhancement models are trained by 200 epochs with a single NVIDIA RTX 3090 GPU.

4.3 Ablation Study

In this section, we evaluate the effectiveness of each part, including PSRB, multiple DIP units, the selection of differentiable filters, and λ_{rec}. The ablation study is performed on 2017MLT, SVT, Real, and the recognition accuracy is calculated by the pre-trained CRNN [3].

Ablation Study on PSRB. We compared the PSRB prediction with its CNN prediction, as shown in Table 2. Benefiting from the ability to extract more textual information from haze images using PSRB, our proposed PSRB prediction learnable parameters are, on average 0.92% higher than the CNN prediction learnable parameters. We used PSRB to provide learnable parameters for the filter in the next ablation experiments.

---

[1] https://github.com/ayumiymk/aster.pytorch.

[2] https://github.com/meijieru/crnn.pytorch.

[3] https://github.com/Canjie-Luo/MORAN_v2.

Table 2. Ablation Study on PSRB.

Method	2017MLT	SVT	Real	average
Baseline	63.49%	52.70%	74.48%	63.55%
CNNPP [35]	70.59%	69.86%	74.74%	71.73%
PSRB	**71.29%**	**71.41%**	**75.26%**	**72.65%**

Table 3. Ablation Study on Multiple DIP.

Method	2017MLT	SVT	Real	average
DIP-1	71.29%	71.41%	75.26%	72.65%
DIP-2	70.56%	67.85%	75.52%	71.31%
DIP-3	70.63%	68.16%	76.28%	71.69%
DIP-4	70.63%	71.26%	76.28%	72.72%
DIP-5	**71.77%**	**72.02%**	**77.57%**	**73.78%**
DIP-6	71.42%	68.93%	76.54%	72.29%

Ablation Study Multiple DIP. A single DIP unit is insufficient for effectively removing various types of haze in different scenes. Therefore, we investigated the combination of multiple DIP units. As shown in Table 3, the average accuracy of DIP-5 is 1.23% higher than DIP-1. We believe that multiple DIP units can effectively remove haze, making them more robust for complex text images. However, when using more DIP units, the performance on average decreased by 1.49%. This phenomenon may be attributed to the fact that an excessive number of DIP units could result in overly enhanced images, which are less conducive to recognition by the recognizer. Therefore, we utilize DIP-5 to conduct the following ablation studies.

Table 4. Ablation Study on balances parameter.

λ_{rec}	2017MLT	SVT	Real	average
0	71.77%	72.02%	**77.57%**	73.78%
0.5	72.68%	**74.03%**	76.54%	**74.41%**
0.1	**72.85%**	72.8%	75.51%	73.72%
0.05	72.60%	73.26%	76.28%	74.04%
0.01	72.31%	72.80%	75.51%	73.54%

Ablation Study on Balances Parameter. We explored the impact on the model from $\{0, 0.5, 0.1, 0.05, 0.01\}$, and the results are shown in Table 4, when the model outperformed all comparable models, compared to when 0.63% improved the average accuracy of CRNN. We believe that when using weights with cross-entropy loss can indeed pave the way for distinguishing confusable characters, thus improving the recognition performance of the recognition network.

Ablation Study on Filter Selection. We evaluated the selection of filters for recognition accuracy using three test datasets. The results are shown in Table 5. By combining the two sets of group filters, the best results were obtained for model C, proving the effectiveness of these filters.

Table 5. Ablation Study on filter selection.

Method	Pixel_wise Filters	Sharpen Filter	2017MLT	SVT	Real	average
A	√		70.92%	71.1%	74.74%	72.25%
B		√	71.13%	70.79%	**76.54%**	72.82%
C	√	√	**72.68%**	**74.03%**	**76.54%**	**74.41%**

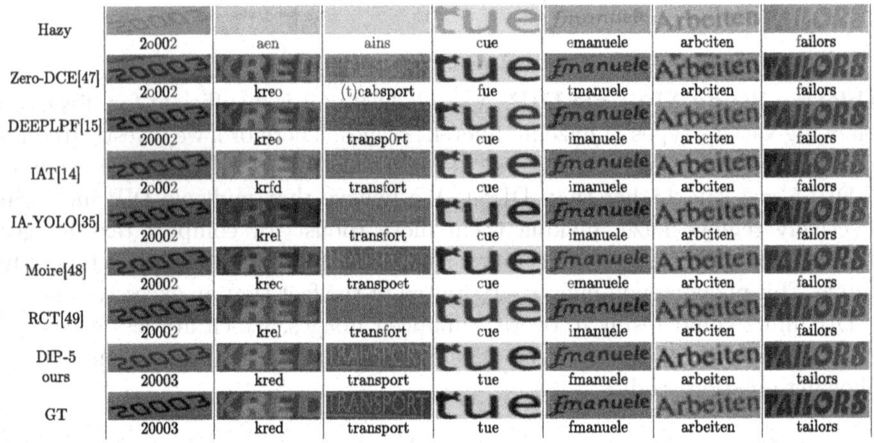

Fig. 4. Visualization results of state-of-the-art image enhancement methods on the three datasets. The character strings under the images are recognition results of CRNN [3]. Those in red denote wrong recognition

4.4 Comparison with State-of-the-Art Image Enhancement Methods

In Table 6, we exemplify the experimental results achieved by six different image enhancement algorithms on the 2017MLT, SVT, Real dataset, including IA-YOLO [35], Moire [48], IAT [14], RCT [49], DEEPLPF [15], and Zero-DCE [47]. Our enhancement method with direct recognition of text images under haze conditions can bring 2.38%, 7.08%, and 9.61% improvement in recognition accuracy for ASTER, MORAN, and CRNN, respectively, compared with the use of CNNPP prediction filter parameters on IA-YOLO used to enhance images.

Our design of PSRB can bring 2.04% improvement in recognition accuracy for ASTER, MORAN, and CRNN, bringing 2.04%, 0.15%, and 1.43% improvement in recognition accuracy, respectively. In addition, our approach yields a performance gain of 2–6% for text recognizers when compared to several mainstream general-purpose image enhancement algorithms. Furthermore, when employing a multiple DIP approach, we still observe a recognition performance gain of 1.25%. Qualitative results are illustrated in Fig. 4. The utilization of PSRB in the sequence context modeling approach, along with the incorporation of multiple DIP units, allows for the retention of more intra and inter-character contextual information.

Table 6. Performance of state-of-the-art image enhancement methods on the three datasets. For better displaying, we calculated the average accuracy. The recognition accuracies are tested by the official released model of ASTER [4], MORAN [46] and CRNN [3].

Method	Accuracy of ASTER [4]				Accuracy of MORAN [46]				Accuracy of CRNN [3]			
	2017MLT	SVT	Real	average	2017MLT	SVT	Real	average	2017MLT	SVT	Real	average
Baseline	74.34%	74.34%	79.38%	76.02%	67.49%	63.37%	84.02%	71.62%	63.49%	52.70%	74.48%	63.55%
Moire [48]	76.59%	77.28%	74.48%	76.11%	74.47%	75.43%	80.92%	76.94%	71.40%	72.02%	73.96%	72.46%
IAT [14]	76.41%	78.36%	68.04%	74.27%	74.31%	74.34%	71.64%	73.43%	70.49%	70.94%	59.53%	66.98%
RCT [49]	77.03%	78.36%	69.84%	75.07%	75.09%	75.89%	77.32%	76.10%	71.67%	73.72%	66.75%	70.71%
DEEPLPF [15]	76.68%	77.28%	62.88%	72.28%	74.74%	75.58%	67.78%	72.70%	71.16%	72.33%	60.05%	67.84%
Zero-DCE [47]	74.93%	76.04%	76.80%	75.92%	72.85%	71.41%	82.47%	75.57%	69.41%	67.54%	71.13%	69.36%
IA-YOLO [35]	76.41%	**78.98%**	73.71%	76.36%	74.72%	75.89%	85.05%	78.55%	70.59%	69.86%	74.74%	71.73%
DIP (ours)	77.43%	77.90%	79.89%	78.40%	75.23%	**76.35%**	84.53%	78.70%	72.36%	71.87%	75.25%	73.16%
DIP-5 (ours)	**77.61%**	78.05%	**80.15%**	**78.60%**	**75.58%**	76.04%	**86.34%**	**79.32%**	**72.68%**	**74.03%**	**76.54%**	**74.41%**

5   Conclusion

This work addresses the problem of text recognition under haze conditions. To the best of our knowledge, this is the first work that addresses text recognition under haze conditions using image enhancement methods. The proposed PSRB prediction filter learns parameters. The pre-training recognition module helps enhance the model's output to facilitate the recognition of text images with good visual effects simultaneously. At the same time, we use a multiple DIP approach to improve the texture details of the text in the images. Our proposed model significantly improves the performance of the text recognition algorithm under haze conditions, which can outperform the six mainstream SOTA image enhancement methods.

References

1. Björklund, T., Fiandrotti, A., Annarumma, M., Francini, G., Magli, E.: Robust license plate recognition using neural networks trained on synthetic images. Pattern Recogn. **93**, 134–146 (2019)

2. Ray, A., et al.: An end-to-end trainable framework for joint optimization of document enhancement and recognition. In: 2019 International Conference on Document Analysis and Recognition (ICDAR), pp. 59–64. IEEE (2019)
3. Shi, B., Bai, X., Yao, C.: An end-to-end trainable neural network for image-based sequence recognition and its application to scene text recognition. IEEE Trans. Pattern Anal. Mach. Intell. **39**(11), 2298–2304 (2016)
4. Shi, B., Yang, M., Wang, X., Lyu, P., Yao, C., Bai, X.: ASTER: an attentional scene text recognizer with flexible rectification. IEEE Trans. Pattern Anal. Mach. Intell. **41**(9), 2035–2048 (2018)
5. Yan, R., Peng, L., Xiao, S., Yao, G.: Primitive representation learning for scene text recognition. In: Proceedings of the IEEE/CVF Conference on Computer Vision and Pattern Recognition, pp. 284–293 (2021)
6. Lu, N., et al.: Master: multi-aspect non-local network for scene text recognition. Pattern Recogn. **117**, 107980 (2021)
7. Hong, M., Xie, Y., Li, C., Qu, Y.: Distilling image dehazing with heterogeneous task imitation. In: Proceedings of the IEEE/CVF Conference on Computer Vision and Pattern Recognition, pp. 3462–3471 (2020)
8. Chen, Z., He, Z., Lu, Z.-M.: DEA-Net: single image dehazing based on detail-enhanced convolution and content-guided attention, arXiv preprint arXiv:2301.04805 (2023)
9. Guo, C.-L., Yan, Q., Anwar, S., Cong, R., Ren, W., Li, C.: Image dehazing transformer with transmission-aware 3d position embedding. In: Proceedings of the IEEE/CVF Conference on Computer Vision and Pattern Recognition, pp. 5812–5820 (2022)
10. Mou, Y., et al.: PlugNet: degradation aware scene text recognition supervised by a pluggable super-resolution unit. In: Vedaldi, A., Bischof, H., Brox, T., Frahm, J.-M. (eds.) ECCV 2020. LNCS, vol. 12360, pp. 158–174. Springer, Cham (2020). https://doi.org/10.1007/978-3-030-58555-6_10
11. Chen, J., Li, B., Xue, X.: Scene text telescope: text-focused scene image super-resolution. In: Proceedings of the IEEE/CVF Conference on Computer Vision and Pattern Recognition, pp. 12 026–12 035 (2021)
12. Ma, J., Guo, S., Zhang, L.: Text prior guided scene text image super-resolution, arXiv preprint arXiv:2106.15368 (2021)
13. Qin, R., Wang, B., Tai, Y.-W.: Scene text image super-resolution via content perceptual loss and criss-cross transformer blocks, arXiv preprint arXiv:2210.06924 (2022)
14. Z. Cui, K. Li, L. Gu, S. Su, P. Gao, Z. Jiang, Y. Qiao, and T. Harada, "Illumination adaptive transformer," arXiv preprint arXiv:2205.14871, 2022
15. Moran, S., Marza, P., McDonagh, S., Parisot, S., Slabaugh, G.: DeepLPF: deep local parametric filters for image enhancement. In: Proceedings of the IEEE/CVF Conference on Computer Vision and Pattern Recognition, pp. 12 826–12 835 (2020)
16. Chen, X., Jin, L., Zhu, Y., Luo, C., Wang, T.: Text recognition in the wild: a survey. ACM Comput. Surv. (CSUR) **54**(2), 1–35 (2021)
17. He, P., Huang, W., Qiao, Y., Loy, C.C., Tang, X.: Reading scene text in deep convolutional sequences. In: Thirtieth AAAI Conference on Artificial Intelligence (2016)
18. Hu, W., Cai, X., Hou, J., Yi, S., Lin, Z.: GTC: guided training of CTC towards efficient and accurate scene text recognition. In: Proceedings of the AAAI Conference on Artificial Intelligence, vol. 34, no. 07, pp. 11 005–11 012 (2020)

19. Graves, A., Fernández, S., Gomez, F., Schmidhuber, J.: Connectionist temporal classification: labelling unsegmented sequence data with recurrent neural networks. In: Proceedings of the 23rd International Conference on Machine Learning, pp. 369–376 (2006)
20. Jaderberg, M., Simonyan, K., Zisserman, A., et al.: Spatial transformer networks. In: Advances in Neural Information Processing Systems, vol. 28 (2015)
21. Li, H., Wang, P., Shen, C., Zhang, G.: Show, attend and read: a simple and strong baseline for irregular text recognition. In: Proceedings of the AAAI Conference on Artificial Intelligence, vol. 33, no. 01, pp. 8610–8617 (2019)
22. Wang, T., et al.: Decoupled attention network for text recognition. In: Proceedings of the AAAI Conference on Artificial Intelligence, vol. 34, no. 07, pp. 12 216–12 224 (2020)
23. Lyu, P., Yang, Z., Leng, X., Wu, X., Li, R., Shen, X.: 2d attentional irregular scene text recognizer, arXiv preprint arXiv:1906.05708 (2019)
24. Lee, J., Park, S., Baek, J., Oh, S.J., Kim, S., Lee, H.: On recognizing texts of arbitrary shapes with 2d self-attention. In: Proceedings of the IEEE/CVF Conference on Computer Vision and Pattern Recognition Workshops, pp. 546–547 (2020)
25. Vaswani, A., et al.: Attention is all you need. In: Advances in Neural Information Processing Systems, vol. 30 (2017)
26. Wang, W., et al.: Scene text image super-resolution in the wild. In: Vedaldi, A., Bischof, H., Brox, T., Frahm, J.-M. (eds.) ECCV 2020. LNCS, vol. 12355, pp. 650–666. Springer, Cham (2020). https://doi.org/10.1007/978-3-030-58607-2_38
27. Chen, J., Yu, H., Ma, J., Li, B., Xue, X.: Text gestalt: stroke-aware scene text image super-resolution. In: Proceedings of the AAAI Conference on Artificial Intelligence, vol. 36, no. 1, pp. 285–293 (2022)
28. Qin, X., Wang, Z., Bai, Y., Xie, X., Jia, H.: FFA-Net: feature fusion attention network for single image dehazing. In: Proceedings of the AAAI Conference on Artificial Intelligence, vol. 34, no. 07, pp. 11 908–11 915 (2020)
29. Qu, Y., Chen, Y., Huang, J., Xie, Y.: Enhanced pix2pix dehazing network. In: Proceedings of the IEEE/CVF Conference on Computer Vision and Pattern Recognition, pp. 8160–8168 (2019)
30. Dong, H., et al.: Multi-scale boosted dehazing network with dense feature fusion. In: Proceedings of the IEEE/CVF Conference on Computer Vision and Pattern Recognition, pp. 2157–2167 (2020)
31. Wu, H., et al.: Contrastive learning for compact single image dehazing. In: Proceedings of the IEEE/CVF Conference on Computer Vision and Pattern Recognition, pp. 10 551–10 560 (2021)
32. Huang, S.-C., Le, T.-H., Jaw, D.-W.: DSNet: joint semantic learning for object detection in inclement weather conditions. IEEE Trans. Pattern Anal. Mach. Intell. **43**(8), 2623–2633 (2020)
33. Kalwar, S., Patel, D., Aanegola, A., Konda, K.R., Garg, S., Krishna, K.M.: GDIP: gated differentiable image processing for object-detection in adverse conditions, arXiv preprint arXiv:2209.14922 (2022)
34. Li, B., Peng, X., Wang, Z., Xu, J., Feng, D.: AOD-Net: all-in-one dehazing network. In: Proceedings of the IEEE International Conference on Computer Vision, pp. 4770–4778 (2017)
35. Liu, W., Ren, G., Yu, R., Guo, S., Zhu, J., Zhang, L.: Image-adaptive yolo for object detection in adverse weather conditions. In: Proceedings of the AAAI Conference on Artificial Intelligence, vol. 36, no. 2, pp. 1792–1800 (2022)

36. Raghunandan, K., et al.: Riesz fractional based model for enhancing license plate detection and recognition. IEEE Trans. Circuits Syst. Video Technol. **28**(9), 2276–2288 (2017)

37. Rahman, M.J., Beauchemin, S.S., Bauer, M.A.: License plate detection and recognition: an empirical study. In: Arai, K., Kapoor, S. (eds.) CVC 2019. AISC, vol. 943, pp. 339–349. Springer, Cham (2020). https://doi.org/10.1007/978-3-030-17795-9_24

38. Osipov, A., et al.: Deep learning method for recognition and classification of images from video recorders in difficult weather conditions. Sustainability **14**(4), 2420 (2022)

39. Rezaei, H., Haghshenas, M., Yasini, M.: Recognizing Persian automobile license plates under adverse rainy conditions. In: 2020 International Conference on Machine Vision and Image Processing (MVIP), pp. 1–8. IEEE (2020)

40. Wang, W., et al.: TextSR: content-aware text super-resolution guided by recognition. arXiv preprint arXiv:1909.07113 (2019)

41. Simonyan, K., Zisserman, A.: Very deep convolutional networks for large-scale image recognition, arXiv preprint arXiv:1409.1556 (2014)

42. Jaderberg, M., Simonyan, K., Vedaldi, A., Zisserman, A.: Synthetic data and artificial neural networks for natural scene text recognition, arXiv preprint arXiv:1406.2227 (2014)

43. Gupta, A., Vedaldi, A., Zisserman, A.: Synthetic data for text localisation in natural images. In: Proceedings of the IEEE Conference on Computer Vision and Pattern Recognition, pp. 2315–2324 (2016)

44. Nayef, N., et al.: ICDAR2017 robust reading challenge on multi-lingual scene text detection and script identification-RRC-MLT. In: 2017 14th IAPR International Conference on Document Analysis and Recognition (ICDAR), vol. 1, pp. 1454–1459. IEEE (2017)

45. Wang, K., Babenko, B., Belongie, S.: End-to-end scene text recognition. In: International Conference on Computer Vision 2011, pp. 1457–1464. IEEE (2011)

46. Luo, C., Jin, L., Sun, Z.: MORAN: a multi-object rectified attention network for scene text recognition. Pattern Recogn. **90**, 109–118 (2019)

47. Guo, C., et al.: Zero-reference deep curve estimation for low-light image enhancement. In: Proceedings of the IEEE/CVF Conference on Computer Vision and Pattern Recognition, pp. 1780–1789 (2020)

48. Sun, Y., Yu, Y., Wang, W.: Moiré photo restoration using multiresolution convolutional neural networks. IEEE Trans. Image Process. **27**(8), 4160–4172 (2018)

49. Kim, H., Choi, S.-M., Kim, C.-S., Koh, Y.J.: Representative color transform for image enhancement. In: Proceedings of the IEEE/CVF International Conference on Computer Vision, pp. 4459–4468 (2021)

Reading Between the Lanes: Text VideoQA on the Road

George Tom[1]([✉]) [ID], Minesh Mathew[1] [ID], Sergi Garcia-Bordils[2,3] [ID],
Dimosthenis Karatzas[2] [ID], and C. V. Jawahar[1] [ID]

[1] Center for Visual Information Technology (CVIT),
IIIT Hyderabad, Hyderabad, India
{george.tom,minesh.mathew}@research.iiit.ac.in, jawahar@iiit.ac.in
[2] Computer Vision Center (CVC), UAB, Barcelona, Spain
{sergi.garcia,dimos}@cvc.uab.cat
[3] AllRead Machine Learning Technologies, Barcelona, Spain

Abstract. Text and signs around roads provide crucial information for drivers, vital for safe navigation and situational awareness. Scene text recognition in motion is a challenging problem, while textual cues typically appear for a short time span, and early detection at a distance is necessary. Systems that exploit such information to assist the driver should not only extract and incorporate visual and textual cues from the video stream but also reason over time. To address this issue, we introduce RoadTextVQA, a new dataset for the task of video question answering (VideoQA) in the context of driver assistance. RoadTextVQA consists of 3,222 driving videos collected from multiple countries, annotated with 10,500 questions, all based on text or road signs present in the driving videos. We assess the performance of state-of-the-art video question answering models on our RoadTextVQA dataset, highlighting the significant potential for improvement in this domain and the usefulness of the dataset in advancing research on in-vehicle support systems and text-aware multimodal question answering. The dataset is available at http://cvit.iiit.ac.in/research/projects/cvit-projects/roadtextvqa.

Keywords: VideoQA · scene text · driving videos

1 Introduction

In this work, we propose a new dataset for Visual Question Answering (VQA) on driving videos, with a focus on questions that require reading text seen on the roads and understanding road signs. Text and road signs provide important information to the driver or a driver assistance system and help to make informed decisions about their route, including how to reach their destination safely and efficiently. Text on roads can also provide directions, such as turn-by-turn directions or the distance to a destination. Road signs can indicate the location of exits, rest stops, and potential hazards, such as road construction or detours.

G. A. Fink et al. (Eds.): ICDAR 2023, LNCS 14192, pp. 137–154, 2023.
https://doi.org/10.1007/978-3-031-41731-3_9

Q: What is the speed limit of the road?   Road Sign based   Answer is present in the video

Ground Truth: [25]   **M4C:** 45   **SINGULARITY:** 40   **GIT:** 45

Q: Which pump company is advertised on the black van on the left?   Text based   Answer is present in the video

Ground Truth: [ketcham]   **M4C:** ketcham   **SINGULARITY:** fedex   **GIT:** hp

Fig. 1. Examples from our RoadTextVQA dataset. The question in the first clip is based on the speed limit road sign, so it is classified as a "road sign based" question. Meanwhile, the question for the clip in the second row draws information from the text on the van, making it a "text based" question. The ground truth answers and the baseline predictions are also presented.

Reading text and understanding road signs is also important for following traffic laws and regulations. Speed limit signs, yield signs, and stop signs provide important information that drivers must follow to ensure their own safety and the safety of others on the road (Fig. 1).

VQA is often dubbed as the Turing test for image/video understanding. The early datasets for VQA on images and videos [2,36,42] largely ignored the need for reading and comprehending text on images and videos, and questions were mostly focus on the visual aspects of the given image or video. For example, questions focused on the type, attributes and names of objects, things or people. However, the text is ubiquitous in outdoor scenes, and this is evident from the fact that nearly 50% of the images in the MS-COCO dataset have text in them [38]. Realizing the importance of reading text in understanding visual scenes, two datasets—Scene text VQA [5] and Text VQA [35] were introduced that focus exclusively on VQA involving scene text in natural images. Two recent works called NewsVideoQA [15], and M4-ViteVQA [22] extend text-based VQA works to videos by proposing VQA tasks that exclusively focus on question-answers that require systems to read the text in the videos.

Similar to these works that focus on text VQA on videos, our work proposes a new dataset where all the questions need to be answered by watching driving videos and reading the text in them. However, in contrast to NewsVideoQA which contains news videos where question-answer pairs are based on video text (born-digital embedded text) appearing on news tickers and headlines, the text in videos in our dataset are scene text. The text in the road or driving videos are subjected to blur, poor contrast, lighting conditions and distortions. Text while driving goes by fast and tends to be heavily occluded. Often, multiple frames needs to be combined to reconstruct the full text, or a good frame with readable text needs to be retrieved. These difficulties made researchers focus on road-text recognition exclusively, and there have been works that focus exclusively on the detection, recognition and tracking of road text videos [12,32]. On the other hand M4-ViteVQA contains varied type of videos such as sports videos, outdoor videos and movie clips. A subset of these videos are driving videos. In contrast, our dataset is exclusively for VQA on driving videos and contains at least three times more questions than in the driving subset of M4-ViteQA. Additionally, questions in our dataset require both reading road text and understanding road signs, while M4-ViteVQA's focus is purely on text-based VQA.

Specifically our contributions are the following:

- We introduce the first large scale dataset for road text and road sign VQA containing 10K+ questions and 3K+ videos.
- We provide a thorough analysis of the dataset and present detailed statistics of videos, questions and answers. We also establish heuristic baselines and upper bounds that help to estimate the difficulty of the problem.
- We evaluate an existing popular VQA model and two SoTA VideoQA models on our dataset and demonstrate that these models fail to perform well on the new dataset since they are not designed to read and reason about text and road signs.

2   Related Work

2.1   VideoQA

In video question answering (VideoQA), the goal is to answer the question in the context of the video. Earlier approaches to VideoQA use LSTM to encode the question and videos [19,27,40,48]. Several datasets have been created in recent years to assist research in the field of video question answering (VideoQA). Large datasets such as MSRVTT-QA [42] contain synthetic generated questions and answers where the questions require only an understanding of the visual scenes. MOVIE-QA [36] and TVQA [24] are based on scenes in movies and TV shows. Castro et al. [7] introduced a dataset with videos from the outside world for video understanding through VideoQA and Video Evidence Selection for interpretability. MOVIE-QA [36], TVQA [24], HowtoVQA69M [44] provide explicit text in the form of subtitles. Multiple-Choice datasets [24,36,43] consist of a predefined set of options for answers. When compared to open-ended datasets, they can be considered limiting in the context of real-world applications. Synthetically generated datasets [7,42,46] contain questions that are generated through processing video descriptions, narration and template questions. MSRVTT-QA [42] exploits the video descriptions for QA creation. HowToVQA69M [44] uses cross-modal supervision and language models to generate question-answer pairs

Table 1. Comparison of RoadTextVQA with existing video question answering datasets. "Text-based" indicates whether the questions require an understanding of the text present in the videos to answer. "Road-based" questions are datasets which are based on the driving domain. "Synthetic questions" are questions that are not manually annotated and depend on automated methods for question-answer generation. Abbreviations used - OE: Open-ended questions, MC: Multiple choice questions.

Dataset	Text-based	Road-based	Synthetic Questions	#Videos	#Questions	QA type
MovieQA [36]	✗	✗	✗	6.7K	6.4K	MC
MSRVTT-QA [42]	✗	✗	✓	10K	243.6K	OE
Activitynet-QA [46]	✗	✗	✓	5.8K	58K	OE
TVQA [24]	✗	✗	✗	21.7K	152.5K	MC
WildQA [7]	✗	✗	✗	0.4K	0.9K	OE
HowtoVQA69M [44]	✗	✗	✓	69M	69MK	OE
SUTD-TrafficQA [43]	✗	✓	✗	10K	62.5K	MC
NewsVideoQA [15]	✓	✗	✗	3K	8.6K	OE
M4-ViteVQA [47]	✓	✗	✗	7.6K	25.1K	OE
RoadTextVQA	✓	✓	✗	3.2K	10.5K	OE

from narrated videos, whereas ActivityNetQA [46] uses template questions to generate the QA pairs. Xu et al. introduced the SUTD-TrafficQA [43] dataset and the Eclipse model for testing systems' ability to reason over complex traffic scenarios. The SUTD-TrafficQA [43] dataset contains multiple-choice questions that are based on different traffic events. RoadTextVQA is an open-ended dataset that deals with questions related to the text information found in road videos or the signs posted along roads. Recent studies [22,23,25,26] on pretraining transformers on other vision and language tasks have shown excellent results for the VideoQA task. Lei et al. [22], in their study, uncovered the bias present in many video question-answering datasets, which only require information from a single frame to answer, and introduced new tasks aimed at training models to answer questions that necessitate the use of temporal information (Table 1).

2.2 VideoQA Involving Video Text

NewsVideoQA [16] and M4-ViteVQA [47] are two recently introduced datasets that include videos with embedded born-digital text and scene text, respectively. Both datasets require an understanding of the text in videos to answer the questions. Embedded text, sometimes called video text in news videos, is often displayed with good contrast and in an easy-to-read style. Scene text in the RoadTextVQA dataset can be challenging to read due to the factors such as occlusion, blur, and perspective distortion. M4-ViteVQA contains videos from different domains, a few of them being shopping, driving, sports, movie and vlogs. The size of RoadTextVQA is more than three times the size driving subset of M4-ViteVQA. Additionally, a subset of questions in RoadTextVQA also requires domain knowledge to answer questions related to road signs. Few recent works [8,41] on vision and language transformers have shown to work well with text-based VQA tasks. Kil et al. [18] introduced PreSTU, a pretraining method that improves text recognition and connects the recognized text with the rest of the image. GIT (GenerativeImage2Text) [41] is a transformer-based model for vision and language tasks with a simple architecture that does not depend on external OCR or object detectors.

2.3 Scene Text VQA

Our work, which focuses on VQA requiring text comprehension within videos, shares similarities with other studies dealing with text in natural images, commonly known as Scene Text VQA. The ST-VQA [5] and TextVQA [35] datasets were the first to incorporate questions requiring understanding textual information from natural images. LoRRa [35] and M4C [14] utilized pointer networks [39] that generate answers from a fixed vocabulary and OCR tokens. In addition, M4C used a multimodal transformer [37] to integrate different modalities. TAP [45] employed a similar architecture to M4C and incorporated a pretraining task based on scene text, improving the model's alignment among the three modalities. Another study, LaTr [4], focused on pretraining on text and layout information from document images and found that incorporating layout information from scanned documents improves the model's understanding of scene text.

3   RoadTextVQA Dataset

This section looks at the data collection and annotation procedure, data analysis, and statistics.

3.1   Data Collection

The videos used in the dataset are taken from the RoadText-3K [12] dataset and YouTube. The RoadText-3K dataset includes 3,000 ten-second road videos that are well-suited for annotation because they have a considerable quantity of text. The RoadText-3K dataset includes videos recorded in the USA, Europe, and India and features text in various languages such as English, Spanish, Catalan, Telugu and Hindi. Each video contains an average of 31 tracks. However, the European subset is excluded from the annotation process for RoadTextVQA as it is dominated by texts in Spanish/Catalan, and the RoadTextVQA is designed specifically for English road-text. In addition to the videos from RoadText-3K, additional dashcam videos were sourced from the YouTube channel J Utah[1]. 252 videos from USA and UK were selected, and clips with a substantial amount of text were further selected by running a text detector over the video frames. Being a free and open-source text detector popular for scene text detection, we went with EasyOCR [17] as our choice of text detector. The RoadText-3K videos

---

[1] https://www.youtube.com/@jutah.

have a resolution of 1280 × 720 with a frame rate of 30 fps. To keep the data consistent, the YouTube clips were downsampled to the same resolution and frame rate of 1280 × 720 at 30 fps.

Individuals who are proficient in the English language were hired to create the question-answer pairs. To ensure the quality of the applicants, an initial training session was conducted, followed by a filtering mechanism in the form of a comprehensive quiz. The quiz was designed to ensure that the question-answer pairs were created by individuals who had a solid grasp of the English language and a good understanding of the task, thereby enabling us to maintain a high standard of quality in the annotations. The annotation process involved two stages, and a specifically designed web-based annotation tool was used. In the initial stage, annotators add the question, answers and timestamp triads for videos shown to them. All the questions have to be based on either some text present in the video or on any road sign. In cases where a question could have multiple answers in a non-ambiguous way, the annotators were given the option to enter several answers. The timestamp is an additional data point which is collected, and it is the aptest point in the video at which the question is answerable. The annotators were instructed to limit the number of questions to

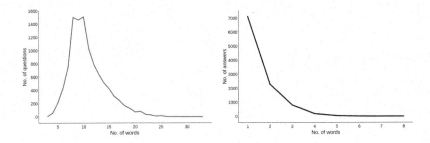

Fig. 2. Distribution of the number of words in the question (left) and answer (right) of RoadTextVQA

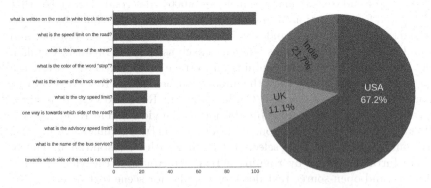

Fig. 3. Top 10 questions in the dataset. **Fig. 4.** Geographical distribution of videos in the RoadTextVQA dataset.

not more than ten per video and to avoid asking any questions related to the
vehicle license plate numbers. If there were no possible questions that could be
asked from the video, then the annotators were given the option to reject it. In
the verification stage, the video and the questions are shown, and the annotators
had to add the answers and the timestamps. We made sure that verification is
done by an annotator different from the one who has annotated it in the first
stage.

Table 2. Comparison of average and maximum question and answer lengths with other
text based video question answering datasets.

Dataset	Average Length		Max Length	
	Question	Answer	Question	Answer
M4-ViteVQA [47]	6.75	1.94	24	26
NewsVideoQA [15]	7.04	2.02	20	19
RoadTextVQA	10.78	1.45	33	8

If the question is incorrect or does not follow the annotation guidelines, it is
flagged and rejected. If for a question, there are common answers in the annotation stage and verification stage, then that question is considered valid. All

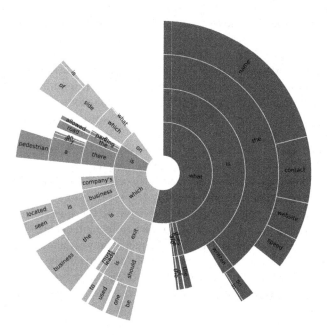

Fig. 5. An analysis of the distribution of questions based on their starting 4 g has
shown that a significant proportion of questions are aimed at obtaining the name and
contact information of businesses located along roads, as well as obtaining the speed
limit for the road.

Fig. 6. The number of occurrences of the answers in RoadTextVQA. The most recurring answer is "right", which makes up about 8% of the answers.

the common answers are considered valid answers to the question. In the verification stage, additional data regarding the question-answers are also collected. The questions are categorically tagged into two distinct classes. Firstly, based on the type of question— text-based or traffic sign-based. The second classification captures whether the answer for a question, i.e., the text that makes up the answer, is present in the video or not.

3.2 Data Statistics and Analysis

The RoadTextVQA dataset contains 3,222 videos and 10,500 question-answer pairs. Among the 3,222 videos, 1,532 videos are taken from the RoadText-3K dataset and the rest are from YouTube. The data is randomly split into 2,557 videos and 8,393 questions in the train set, 329 videos and 1,052 questions in the test, and 336 videos and 1,055 questions in the validation set.

The videos for the test and validation sets were randomly chosen from the RoadText-3K split, as it has ground truth annotations for text tracking. Methods that use OCR data can take advantage of the accurate annotations provided by RoadText-3K.

We present statistics related to the questions in RoadTextVQA through Fig. 2, and Fig. 3. Figure 3 shows the most frequent questions and their frequencies. "What is written on the road with white block letters?" is the most recurrent, followed by questions regarding the speed limits on the roads.

Figure 5 provides a comprehensive overview of the question distribution in RoadTextVQA, with the majority of the questions being centred around details of shops located along the road. Figure 2 depicts the word count in the questions and answers, respectively. The average number of words in the questions in Road-TextVQA is 10.8, while the average number in the answers is 1.45. The average number of words in questions is much higher when compared to other text-based VideoQA datasets, as seen in Table 2. The percentage of unique questions stands at 86.6%, while the percentage of unique answers is 40.7%. Figure 6 shows the top 30 answers and the number of occurrences. Figure 7, in the form of a word cloud, illustrates the most frequently occurring answers and OCR tokens. The most popular answers are "right", "left", "yes", and "no". The most prevalent

Fig. 7. A visual representation of word frequency in the form of a word cloud, depicting the distribution of words in answers (left) and the distribution of OCR tokens from the videos (right).

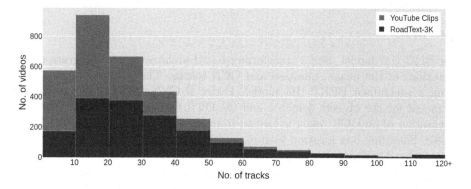

Fig. 8. Distribution of number of videos vs number of tracks.

OCR tokens in the videos are "stop", "only", and "one way". The distribution of the videos in the dataset based on the geographic location where it was captured is shown in Fig. 4. More than two-thirds of the videos in the dataset are captured from roads in the USA.

The majority of questions are grounded on text seen in the video (61.8%), and the rest are based on road signs. Road signs can also contain text, such as speed limit signs or interchange exit signs. 68% of questions have answers that can be found within the text present in the video, while the remaining 32% of questions require an answer that is not a text present in the video.

4 Baselines

This section presents details of the baselines we evaluate on the proposed Road-TextVQA dataset.

4.1   Heuristic Baselines and Upper Bounds

We evaluate several heuristic baselines and upper bounds on the dataset. These heuristics and upper bounds are similar to those used in other VQA benchmarks, such as TextVQA [35] and DocVQA [30]. The following heuristic baselines are evaluated: (i) **Random Answer:** performance when answers to questions are randomly selected from the train split. (ii) **Random OCR token:** performance when a random OCR token from the video is picked as the answer. (iii) **Majority Answer:** performance when the most common answer in the train split is considered as the answer for all the questions. The following upper bounds are evaluated (i) **Vocab UB:** the upper bound on predicting the correct answer if it is present in the vocabulary of all the answers from the train split. (ii) **OCR UB:** the upper bound on performance if the answer corresponds to an OCR token present in the video. (iii) **Vocab UB + OCR UB:** this metric reflects the proportion of questions for which answers can be found in the vocabulary or the OCR transcriptions of the video.

4.2   M4C

The M4C [14] model uses a transformer-based architecture to integrate representations of the image, question and OCR tokens. The question is embedded using a pretrained BERT [10] model. Faster R-CNN [33] visual features are extracted for the objects detected and the OCR tokens in the image. The representation of an OCR token is formed from the FastText [6] vector, PHOC [1] vector, bounding box location feature, and Faster R-CNN feature of the token. A multi-head self-attention mechanism in transformers is employed, enabling all entities to interact with each other and model inter- and intra-modal relationships uniformly using the same set of transformer parameters. During answer prediction, the M4C model employs an iterative, auto-regressive decoder that predicts one word at a time. The decoder can use either a fixed vocabulary or the OCR tokens detected in the image to generate the answer.

4.3   Singularity

The architecture of SINGULARITY [22] is made up of three major components: a vision encoder using ViT [11], a language encoder utilizing BERT [10], and a multi-modal encoder using a transformer encoder [37]. The multi-modal encoder uses cross-attention to collect information from visual representations using text as the key. Each video or image is paired with its corresponding caption during the pretraining phase, and the model is trained to align the vision and text representations using three losses (i) Vision-Text Contrastive: a contrastive loss which aligns the representations of vision and language encoders, (ii) Masked Language Modeling [10]: masked tokens are predicted (iii) Vision-Text Matching: using the multi-modal encoder, predict the matching score of a vision-text pair. We use the SINGULARITY-temporal model, which is pretrained on 17M vision caption pairs [3,9,21,31,34,38]. The SINGULARITY-temporal model contains a two-layer temporal encoder that feeds its outputs into the multi-modal encoder. SINGULARITY-temporal makes use of two new datasets named SSv2-Template Retrieval, and SSv2-Label Retrieval created from the action recognition dataset Something-Something v2 (SSv2) [13]. The pretraining is a video retrieval task using text queries. An additional multi-modal decoder is added for open-ended QA tasks and is initialised from the pretrained multi-modal encoder, which takes the multi-modal encoder's output as input and generates answer text with [CLS] as the start token.

4.4 GenerativeImage2Text

GIT (GenerativeImage2Text) [41] is a transformer-based architecture aimed at unifying all vision-language tasks using a simple architecture pretrained on 0.8 billion image text pairs. GIT consists of an image encoder and a text decoder and is pretrained on a large dataset of image text pairs. The image encoder is a Swin-like [28] transformer based on the contrastive pretrained model, which eliminates the need for other object detectors or OCR. As for the text decoder, the GIT uses a transformer with a self-attention and feed-forward layer to generate text output. The visual features and the text embeddings are concatenated and used as inputs to the decoder. GIT is able to gradually learn how to read the scene text with large-scale pretraining and hence achieves SoTA performance on scene-text-related VQA tasks such as ST-VQA. For video question answering, GIT employs a method of selecting multiple frames from the video and separately embeds each frame with a learnable temporal embedding which is initialized as zeros, and the image features are concatenated and used similarly to the image representation. The question and the correct answer are combined and used in the sense of a special caption, and the language model loss is computed solely on the answer and the [EOS] token.

5 Experiments and Results

This section covers the evaluation metrics, the experimental setup, and the experiment results.

5.1 Experimental Setup

Evaluation Metrics. We use two evaluation metrics to evaluate the model's performance: Average Normalized Levenshtein Similarity (ANLS) [5] and Accuracy (Acc. (%)). The Accuracy metric calculates the percentage of questions where the predicted answer exactly matches any of the target answers. ANLS, on the other hand, does not award a zero score for all predictions that do not match the ground truth string exactly. The score was originally proposed to act softly on cases where the predicted answer differs slightly from the actual. ANLS measures a similarity (based on the Levenshtein distance) between the prediction and ground truth and normalizes it as a score in the range $[0, 1]$. If the score is less than 0.5, the final ANLS score for the prediction is set to zero.

OCR Transcriptions. The ground truth annotations were utilized for the videos in the RoadText-3K set, while for the remaining videos, the OCR transcriptions were sourced using the Google Cloud Video Intelligence API. Both RoadText-3K ground truth annotations, and the Google API provide text transcriptions at the line level. We use the line-level text transcriptions as the OCR tokens for the calculation of OCR upper bounds and OCR-based heuristics as given in the Table 3. When a text track gets cut off from the frame or partially occluded by other objects in a video, the Google Cloud Video Intelligence API treats it as a new track, whereas RoadText-3K annotations ignore the partially occluded tracks. This is why in the Fig. 8, the number of videos vs the number of tracks is a bit inflated for the YouTube clips when compared to RoadText-3K clips.

Experimental Setup for M4C. The M4C [14] model is trained using the official implementation, and the training parameters and implementation details remain consistent with those used in the original paper. We used a fixed vocabulary of size 3926 generated from the train set. The training data consists of image question-answer pairs where the image selected for training is the one on which the questions are based, specifically the timestamp frame. After training, the model is evaluated using two approaches. Firstly, it is tested on the timestamp QA pairs of the test set, and secondly, it is evaluated on the video level by sampling ten frames from the respective video for each QA pair and obtaining the model prediction for every frame individually. The final answer is determined by taking the most common answer from the ten individual frame predictions.

Experimental Setup for SINGULARITY. We fine-tuned the pretrained SINGULARITY-temporal 17M model on four NVIDIA Geforce RTX 2080 Ti. The fine-tuning process was run for 20 epochs with a batch size of 16, starting with an initial learning rate of $1e-5$ and increasing linearly in the first half epoch, followed by cosine decay [29] to $1e-6$. The other parameters used for training are the same as the official implementation. The video frames were resized to 224×224, and a single frame with random resize, crop and flip augmentations was utilised during training, whereas 12 frames were used during testing. Additionally, we fine-tuned the SINGULARITY model, which has been pretrained on the MSRVTT-QA [42] dataset.

Experimental Setup for GIT. The training process for GIT was carried out using a single Tesla T4 GPU for 20 epochs with a batch size of 2. We use an Adam [20] optimizer with an initial learning rate starting at $1e-5$ and gradually decreasing to $1e-6$ through the use of cosine decay. The GIT model was trained using the official VideoQA configuration used for MSRVTT-QA training. We fine-tuned the pretrained GIT-large model on our dataset, using six frames that were evenly spaced as inputs during both training and testing. In addition, we further fine-tuned the GIT model that was pretrained on the MSRVTT-QA [42] dataset.

5.2 Results

Heuristic baselines and upper bound results are presented in the Table 3. The heuristic baselines yield very low accuracy, which indicates the absence of any bias due to the repetition of answers.

Table 3. Performance of various heuristic baselines and upper bounds that are commonly evaluated on text-based VQA datasets.

Baseline	Test	
	ANLS	Acc. (%)
Random Answer	0.09	0
Random OCR token	3.20	1.98
Majority Answer	–	3.49
Vocab UB	–	59.26
OCR UB	–	36.67
Vocab + OCR UB	–	76.18

Random OCR heuristic gives close to 2% accuracy, meaning that there is enough text present in the video that selecting a random OCR from the video will not yield high accuracy. The OCR upper bound is 36.6% which is low when compared to the percentage of questions which have the answers present in the video. The low OCR UB can be attributed to how the text detection and how ground truth annotation is done. The response to a question may be split into multiple lines within the video, leading to the representation of the answer as separate tokens in the OCR output. This happens because the annotations in the OCR process were carried out on a line level. From the upper bound result of Vocab + OCR UB, we can see that more than three-quarters of the answers are present in either the vocabulary or in the OCR tokens of the video.

The results on M4C are shown on Table 4. The frame level results, where we evaluate on the timestamp frame, show an accuracy of 38.20% and the video level results, where we evaluate on ten frames, give an accuracy of 28.92%. The results show that answering the question is still a challenging task, even when we reduce the complexity of the problem by providing the aptest frame for answering the question and ground truth OCR tokens.

Table 4. Performance of RoadTextVQA on M4C. Abbreviations- TB: text-based questions, RSB: road sign-based questions, AP: questions where the answer is present in the video, ANP: questions where the answer is not present in the video.

Test Frames	TB		RSB		AP		ANP		All	
	ANLS	Acc. (%)	ANLS	Acc. (%)	ANLS	Acc. (%)	ANLS	Acc. (%)	ANLS	Acc. (%)
1	35.28	29.27	55.49	49.46	37.55	29.70	63.85	63.19	44.22	38.20
10	23.92	21.48	42.83	38.32	20.38	15.96	67.12	66.91	32.27	28.92

Table 5. Performance of RoadTextVQA on SINGULARITY and GIT. Abbreviations-TB: text-based questions, RSB: road sign-based questions, AP: questions where the answer is present in the video, ANP: questions where the answer is not present in the video.

Method	Pretrain Data	TB		RSB		AP		ANP		All	
		ANLS	Acc. (%)	ANLS	Acc. (%)	ANLS	Acc. (%)	ANLS	Acc. (%)	ANLS	Acc. (%)
SINGULARITY	–	15.38	14.04	45.29	33.04	17.36	9.22	61.71	61.33	28.62	22.45
SINGULARITY	MSVRTT-QA	17.25	15.22	47.84	36.46	19.50	11.50	63.98	63.19	30.79	24.62
GIT	–	18.09	14.38	50.36	39.82	20.98	12.16	65.65	65.05	32.34	25.61
GIT	MSRVTT-QA	22.61	19.62	51.20	42.18	23.40	15.96	69.93	69.51	35.23	29.58

Q: About what building does the yellow sign board and the writing on the road warn about?
Ground Truth: [school]
M4C: school
SINGULARITY: school
GIT: school

Q: What is the name of the fuel retail outlet on the right?
Ground Truth: [valero]
M4C: valero
SINGULARITY: mcdonald's
GIT: exxon

Q: Which is the airport that one can reach if they go straight ahead from the traffic intersection?
Ground Truth: [rgi airport]
M4C: mehd patnam rgi airport
SINGULARITY: terminal
GIT: kennedy airport

Fig. 9. Qualitative results showing predictions of M4C, SINGULARITY and GIT. The correct predictions are highlighted in green, whereas the incorrect ones are highlighted in red. (Color figure online)

We show the results after fine-tuning on SINGULARITY and GIT in Table 5. The accuracy of the questions requiring answers to be extracted from the video (AP) is comparatively lower, while the accuracy of the questions where the answer is not present in the video is comparatively higher. Compared to AP, ANP is less complex to answer because it involves a fixed set of answers. In contrast, AP requires dynamic extraction from OCR tokens, resulting in the ANP set having better accuracy than AP. Additionally, fine-tuning the model that has been pretrained on the MSRVTT-QA dataset shows improvement in accuracy across all categories (TB, RSB, AP, and ANP).

Fine-tuning GIT results in better performance compared to SINGULARITY. GIT also shows a similar trend when fine-tuned on pretrained MSRVTT-QA dataset. The "answer is present in the videos (AP)" subset has an improvement of 3.9% in accuracy when compared with SINGULARITY, whereas the "answer is not present (ANP)" in the videos subset has a gain of 6.3%. M4C tested on a single frame shows better results compared to VideoQA models. This can be attributed to the fact that we explicitly provide the OCR tokens and the correct frame on which the question is framed to the model. M4C tested on ten frames gives comparable results to GIT.

We show some of the qualitative results in Fig. 9. As the complexity of the scene and the obscurity of the scene text increase, it becomes more and more difficult for the model to predict the correct answer. VideoQA baselines achieve better results on questions that do not require the extraction of answers from the video.

6   Conclusions

We introduce RoadTextVQA, a new Video Question Answering dataset where the questions are grounded on the text and road signs present in the road videos. Our findings from the baseline models' performance indicate a need for improvement in existing VideoQA approaches for text-aware multimodal question answering.

Future work can involve augmenting the dataset by incorporating videos obtained from diverse global locales. Currently, there are recurrent questions and answers due to repeating elements in the videos. Including videos from various locations broadens the diversity of the dataset by providing a more comprehensive range of questions and answers and minimizes any biases within the dataset. To our best knowledge, currently, there are no Visual Question Answering models that explicitly incorporate road signs. Models can integrate road signs as an additional input or pretrain on road sign-description pairs to enhance their ability to respond to questions that require domain knowledge.

We believe this work would encourage researchers to develop better models that incorporate scene text and road signs and are resilient to the challenges posed by driving videos. Additionally, drive further research in the area of scene text VideoQA and the development of advanced in-vehicle support systems.

Acknowledgements. This work has been supported by IHub-Data at IIIT-Hyderabad, and grants PDC2021-121512-I00, and PID2020-116298GB-I00 funded by MCIN/AEI/10.13039/501100011033 and the European Union NextGenerationEU/PRTR.

References

1. Almazán, J., Gordo, A., Fornés, A., Valveny, E.: Word spotting and recognition with embedded attributes. IEEE Trans. Pattern Anal. Mach. Intell. **36**(12), 2552–2566 (2014)
2. Antol, S., et al.: VQA: visual question answering. In: ICCV (2015)
3. Bain, M., Nagrani, A., Varol, G., Zisserman, A.: Frozen in time: a joint video and image encoder for end-to-end retrieval. In: Proceedings of the IEEE/CVF International Conference on Computer Vision, pp. 1728–1738 (2021)

4. Biten, A.F., Litman, R., Xie, Y., Appalaraju, S., Manmatha, R.: LaTr: layout-aware transformer for scene-text VQA. In: Proceedings of the IEEE/CVF Conference on Computer Vision and Pattern Recognition, pp. 16548–16558 (2022)

5. Biten, A.F., et al.: Scene text visual question answering. In: Proceedings of the IEEE/CVF International Conference on Computer Vision (ICCV) (2019)

6. Bojanowski, P., Grave, E., Joulin, A., Mikolov, T.: Enriching word vectors with subword information. Trans. Assoc. Comput. Linguist. **5**, 135–146 (2017)

7. Castro, S., Deng, N., Huang, P., Burzo, M., Mihalcea, R.: In-the-wild video question answering. In: Proceedings of the 29th International Conference on Computational Linguistics, pp. 5613–5635 (2022)

8. Chen, X., et al.: PaLi: a jointly-scaled multilingual language-image model. arXiv preprint arXiv:2209.06794 (2022)

9. Chen, Y.-C., et al.: UNITER: UNiversal Image-TExt representation learning. In: Vedaldi, A., Bischof, H., Brox, T., Frahm, J.-M. (eds.) ECCV 2020. LNCS, vol. 12375, pp. 104–120. Springer, Cham (2020). https://doi.org/10.1007/978-3-030-58577-8_7

10. Devlin, J., Chang, M.W., Lee, K., Toutanova, K.: BERT: pre-training of deep bidirectional transformers for language understanding. arXiv preprint arXiv:1810.04805 (2018)

11. Dosovitskiy, A., et al.: An image is worth 16x16 words: transformers for image recognition at scale. In: ICLR (2021)

12. Garcia-Bordils, S., et al.: Read while you drive - multilingual text tracking on the road. In: Uchida, S., Barney, E., Eglin, V. (eds.) DAS 2022. LNCS, pp. 756–770. Springer, Cham (2022). https://doi.org/10.1007/978-3-031-06555-2_5110.1007/978-3-031-06555-2_51

13. Goyal, R., et al.: The "something something" video database for learning and evaluating visual common sense. In: Proceedings of the IEEE International Conference on Computer Vision, pp. 5842–5850 (2017)

14. Hu, R., Singh, A., Darrell, T., Rohrbach, M.: Iterative answer prediction with pointer-augmented multimodal transformers for textVQA. In: Proceedings of the IEEE/CVF Conference on Computer Vision and Pattern Recognition, pp. 9992–10002 (2020)

15. Jahagirdar, S., Mathew, M., Karatzas, D., Jawahar, C.V.: Watching the news: towards videoQA models that can read (2022)

16. Jahagirdar, S., Mathew, M., Karatzas, D., Jawahar, C.: Watching the news: towards videoQA models that can read. In: Proceedings of the IEEE/CVF Winter Conference on Applications of Computer Vision, pp. 4441–4450 (2023)

17. JaidedAI: EasyOCR. https://github.com/JaidedAI/EasyOCR

18. Kil, J., et al.: PreSTU: pre-training for scene-text understanding. arXiv preprint arXiv:2209.05534 (2022)

19. Kim, J., Ma, M., Pham, T., Kim, K., Yoo, C.D.: Modality shifting attention network for multi-modal video question answering. In: Proceedings of the IEEE/CVF Conference on Computer Vision and Pattern Recognition, pp. 10106–10115 (2020)

20. Kingma, D.P., Ba, J.: Adam: a method for stochastic optimization. In: Bengio, Y., LeCun, Y. (eds.) 3rd International Conference on Learning Representations, ICLR 2015, San Diego, CA, USA, 7–9 May 2015, Conference Track Proceedings (2015)

21. Krishna, R., et al.: Visual genome: connecting language and vision using crowd-sourced dense image annotations. Int. J. Comput. Vision **123**, 32–73 (2017)

22. Lei, J., Berg, T.L., Bansal, M.: Revealing single frame bias for video-and-language learning. arXiv preprint arXiv:2206.03428 (2022)

23. Lei, J., et al.: Less is more: Clipbert for video-and-language learning via sparse sampling. In: Proceedings of the IEEE/CVF Conference on Computer Vision and Pattern Recognition, pp. 7331–7341 (2021)

24. Lei, J., Yu, L., Bansal, M., Berg, T.L.: TVQA: localized, compositional video question answering. In: EMNLP (2018)

25. Li, J., Li, D., Xiong, C., Hoi, S.: BLIP: bootstrapping language-image pre-training for unified vision-language understanding and generation. In: International Conference on Machine Learning, pp. 12888–12900. PMLR (2022)

26. Li, L., Chen, Y.C., Cheng, Y., Gan, Z., Yu, L., Liu, J.: HERO: hierarchical encoder for video+ language omni-representation pre-training. In: Proceedings of the 2020 Conference on Empirical Methods in Natural Language Processing (EMNLP), pp. 2046–2065 (2020)

27. Li, X., et al.: Beyond RNNs: positional self-attention with co-attention for video question answering. In: Proceedings of the AAAI Conference on Artificial Intelligence (2019)

28. Liu, Z., et al.: Swin transformer: hierarchical vision transformer using shifted windows. In: Proceedings of the IEEE/CVF International Conference on Computer Vision (ICCV) (2021)

29. Loshchilov, I., Hutter, F.: SGDR: stochastic gradient descent with warm restarts. arXiv preprint arXiv:1608.03983 (2016)

30. Mathew, M., Karatzas, D., Jawahar, C.: DocVQA: a dataset for VQA on document images. In: WACV, pp. 2200–2209 (2021)

31. Ordonez, V., Kulkarni, G., Berg, T.: Im2Text: describing images using 1 million captioned photographs. In: Advances in Neural Information Processing Systems, vol. 24 (2011)

32. Reddy, S., Mathew, M., Gomez, L., Rusinol, M., Karatzas, D., Jawahar, C.: RoadText-1K: text detection & recognition dataset for driving videos. In: 2020 IEEE International Conference on Robotics and Automation (ICRA), pp. 11074–11080. IEEE (2020)

33. Ren, S., He, K., Girshick, R., Sun, J.: Faster R-CNN: towards real-time object detection with region proposal networks. In: Advances in Neural Information Processing Systems, vol. 28 (2015)

34. Sharma, P., Ding, N., Goodman, S., Soricut, R.: Conceptual captions: a cleaned, hypernymed, image alt-text dataset for automatic image captioning. In: Proceedings of the 56th Annual Meeting of the Association for Computational Linguistics (Volume 1: Long Papers), pp. 2556–2565 (2018)

35. Singh, A., et al.: Towards VQA models that can read. In: Proceedings of the IEEE Conference on Computer Vision and Pattern Recognition, pp. 8317–8326 (2019)

36. Tapaswi, M., Zhu, Y., Stiefelhagen, R., Torralba, A., Urtasun, R., Fidler, S.: MovieQA: understanding stories in movies through question-answering. In: Proceedings of the IEEE Conference on Computer Vision and Pattern Recognition, pp. 4631–4640 (2016)

37. Vaswani, A., et al.: Attention is all you need. In: Advances in Neural Information Processing Systems, vol. 30 (2017)

38. Veit, A., Matera, T., Neumann, L., Matas, J., Belongie, S.: COCO-text: dataset and benchmark for text detection and recognition in natural images. In: arXiv preprint arXiv:1601.07140 (2016)

39. Vinyals, O., Fortunato, M., Jaitly, N.: Pointer networks. In: Advances in Neural Information Processing Systems, vol. 28 (2015)

40. Wang, B., Xu, Y., Han, Y., Hong, R.: Movie question answering: remembering the textual cues for layered visual contents. In: Proceedings of the AAAI Conference on Artificial Intelligence (2018)
41. Wang, J., et al.: GIT: a generative image-to-text transformer for vision and language. arXiv preprint arXiv:2205.14100 (2022)
42. Xu, D., et al.: Video question answering via gradually refined attention over appearance and motion. In: Proceedings of the 25th ACM International Conference on Multimedia, pp. 1645–1653 (2017)
43. Xu, L., Huang, H., Liu, J.: SUTD-trafficQA: a question answering benchmark and an efficient network for video reasoning over traffic events. In: Proceedings of the IEEE/CVF Conference on Computer Vision and Pattern Recognition, pp. 9878–9888 (2021)
44. Yang, A., Miech, A., Sivic, J., Laptev, I., Schmid, C.: Just ask: learning to answer questions from millions of narrated videos. In: Proceedings of the IEEE/CVF International Conference on Computer Vision, pp. 1686–1697 (2021)
45. Yang, Z., et al.: TAP: text-aware pre-training for text-VQA and text-caption. In: Proceedings of the IEEE/CVF Conference on Computer Vision and Pattern Recognition, pp. 8751–8761 (2021)
46. Yu, Z., et al.: ActivityNet-QA: a dataset for understanding complex web videos via question answering. In: AAAI, pp. 9127–9134 (2019)
47. Zhao, M., et al.: Towards video text visual question answering: benchmark and baseline. In: Thirty-sixth Conference on Neural Information Processing Systems Datasets and Benchmarks Track (2022)
48. Zhao, Z., Jiang, X., Cai, D., Xiao, J., He, X., Pu, S.: Multi-turn video question answering via multi-stream hierarchical attention context network. In: IJCAI, vol. 2018, p. 27th (2018)

TPFNet: A Novel Text In-painting Transformer for Text Removal

Onkar Susladkar[1] , Dhruv Makwana[1] , Gayatri Deshmukh[1] ,
Sparsh Mittal[2(✉)] , R. Sai Chandra Teja[1] , and Rekha Singhal[3]

[1] Bengaluru, India
[2] IIT Roorkee, Roorkee, India
`sparshfec@iitr.ac.in`
[3] TCS Research, Mumbai, India
`rekha.singhal@tcs.com`

Abstract. Text erasure from an image is helpful for various tasks such as image editing and privacy preservation. We present TPFNet, a novel one-stage network for text removal from images. TPFNet has two parts: feature synthesis and image generation. Since noise can be more effectively removed from low-resolution images, part 1 operates on low-resolution images. Part 1 uses PVT or EfficientNet-B6 as the encoder. Further, we use a novel multi-headed decoder that generates a high-pass filtered image and a segmentation map, along with a text-free image. The segmentation branch helps locate the text precisely, and the high-pass branch helps in learning the image structure. Part 2 uses the features learned in part 1 to predict a high-resolution text-free image. To precisely locate the text, TPFNet employs an adversarial loss that is conditional on the segmentation map rather than the input image. On Oxford, SCUT, SCUT-EnsText and ICDAR2013 datasets, TPFNet outperforms recent networks on nearly all the metrics. E.g., on Oxford dataset, TPFNet has a PSNR (higher is better) of 44.2 and a text-detection precision (lower is better) of 39.0, compared to MTRNet++'s PSNR of 33.7 and precision of 50.4. The source code can be obtained from https://github.com/CandleLabAI/TPFNet.

1 Introduction

Recent years have seen phenomenal growth in text recognition from images [7,17,19]. Real-life images also contain private or sensitive information such as addresses, cellphone numbers, and other personally identifiable information. Automatic text recognition engines can collect this information and use it for malicious purposes such as marketing, privacy breaches, and identity theft. Text erasure refers to erasing only the text area in the image without changing the pixel values of other image regions. Thus, text erasure is helpful for many applications, such as privacy protection, autonomous driving, support systems for the visually impaired, information provision systems in exhibition halls, and

O. Susladkar, D. Makwana, G. Deshmukh and R. Sai Chandra Teja—Independent Researcher.

translation guidance systems for foreigners. For example, after erasing an advertisement text in one language (e.g., English), the text in another language (e.g., Japanese) can be inserted.

However, erasing the text from images presents unique challenges. Text in images does not usually have any border lines or clear distinction from the background. It is often difficult to utilize border or background color information to detect the text area. The border detection may fail if the background color and text color are the same. A technique is needed to recognize the text at all the images' locations. If the image is photographed at an inclined angle, overlapping adjacent text characters may make text recognition difficult. Further, text erasure techniques must deal with variations in background, texture, format, lighting conditions, font, and layout.

The previous works have proposed both two-stage and one-stage (end-to-end) networks. The two-stage networks (e.g., [23]) first ascertain the text location and then use a segmented mask to remove the text. However, their efficacy depends on the text-detection step. Furthermore, due to the need to train two separate networks, they require a complex training strategy. Some text extraction techniques first recognize individual characters and then connect them to form words. These techniques assume that the character sizes are uniform or make other assumptions about text position. The one-stage networks take an input image and directly produce a final text-free image. However, these techniques have limited efficacy in restoring complex backgrounds. Also, they may lead to blurred, corrupted or incomplete images or images with artifacts.

Some previous techniques [22,23] can perform text-erasure only inside the region shown by a word/character-level mask. On using a mask, the model need not detect the text and can just focus on inpainting. Hence, use of mask improves the efficacy of text-removal, or in other words, it is more challenging to achieve good results without mask. In fact, without a mask, these techniques do not converge quickly during training [23]. Overall, there is a need of novel, effective text erasure techniques.

Contributions: In this paper, we propose TPFNet, a deep-learning model for end-to-end text erasure. As such, TPFNet falls into the category of one-stage text removal networks. Our contributions can be summarized as follows:

1. We note that it is simpler to eliminate noise from a low-resolution image than from a high-resolution one. Based on this, the TPFNet network has two parts: feature synthesis and image generation. In part 1, the network learns the features and creates a low-resolution text-free image, and in part 2, the network uses the learned features to predict a high-resolution text-free image.
2. The generator in part 1 uses a pretrained Pyramidal Vision Transformer (or a similar network such as EfficientNet-B6) as the encoder (Sect. 3.1) and a novel multi-branch decoder. Specifically, the decoder (Sect. 3.2) has three branches that predict a high-pass filtered image, a segmentation map, and an image devoid of text. Each of these outputs assists the model in producing accurate results. Specifically, the high-pass filter branch helps in learning and

regaining the structural knowledge of the image. The segmentation branch helps in precisely locating the text.

3. Unlike other conditional GAN architectures, TPFNet employs an adversarial loss conditional on the segmentation map rather than the input image. This allows TPFNet to pinpoint the text within an image (Sect. 3.4).

4. We rigorously evaluate our network on the Oxford, SCUT, SCUT-EnsText and ICDAR2013 datasets (Sect. 4). On all these datasets, TPFNet outperforms previous works on nearly all the metrics (Sect. 5.1). For example, on Oxford dataset, TPFNet has a PSNR (higher is better) of 44.2 and a text-detection precision (lower is better) of 39.0, compared to the best previous technique (MTRNet++), which has a PSNR of 33.7 and precision of 50.4. TPFNet achieves state-of-art results even without the use of masks and as we show in Table 5, use of mask further boosts the efficacy of TPFNet. Qualitative results confirm these findings (Sect. 5.2). We also evaluate our model on unseen images taken from the internet and observe that our model is effective on these images.

The ablation results provide further insights into the importance of each component (Sect. 6). With a PVT backbone, TPFNet has 59.8M parameters and a throughput of 14 frames per second. Changing the backbone and precision helps exercise a tradeoff between image quality and throughput/model size. TPFNet can erase text in different languages, font-sizes and even slanted text. https:// youtu.be/nkYF0GlZgi8 shows a sample video of TPFNet results.

2   Related Work

Text erasure, a specific case of image inpainting, requires two subtasks: (1) positioning the text area and (2) erasing the text. The initial text removal methods [9,21,24] are two-stage procedures that rely on either standard text detection or inpainting techniques. Image inpainting is made possible in a single stage [14] due to the advances in deep learning. Zhang et al. [28] propose an end-to-end word erasure network, which combines two subtasks. The discriminator and various loss functions guide the learning of the generator. In order to enable the network to perceive the text location better, Liu et al. [11] further introduce a mask branch for learning. Tursun et al. [23] introduce text segmentation results in the input so that the network can more accurately perceive the position information of the text area. Bian et al. [1] achieve specific text stroke detection and stroke removal through a cascading structure. However, these methods need to know the exact location of the text area in advance. Tursun et al. [22] introduce a fine-tuning sub-network to reduce the network's dependence on the input location information, thereby achieving a more robust text erasure algorithm. They use a coarse mask to perform coarse-inpainting and further refine these results

Later efforts [5,12,26] enhance image inpainting by introducing new losses, such as content loss and style loss, in the neural network. These losses are measured using classification models pre-trained on the ImageNet dataset. Yu et al. [27] showed potential for inpainting using adversarial losses acquired from a generative adversarial network (GAN). Recent work [6,16] also demonstrates the benefits of utilizing both a pre-trained classification network and a newly trained discriminator for image inpainting.

Nakamura et al. [15] propose SceneTextEraser, which is the first one-stage text removal approach. It is a patch-based auto-encoder featuring skip connections. By alternating textual and non-textual patches, they successfully train the network. Because networks are trained with just pixel intensity-based losses, the inpainting outputs of the early one-stage technique are fuzzy and lack detail. EnsNet [28] uses a local-aware discriminator and four improved loss functions to keep the erased text uniform. Liu et al. [11] introduce EraseNet, which has a coarse-to-fine architecture and an extra segmentation head to assist with text localization. Compared to EraseNet, MTRNet++ [22] splits the encoding of the image content and the text mask into two distinct streams. Liu et al. [10] develop a Local-global Content Modeling (LGCM) block using CNNs and Transformer-Encoder to construct long-term relationships among pixels; this improves the efficacy of text removal. Table 1 summarizes key ideas of related works.

Table 1. Summary of related works

	Model	Loss-function	Dataset
STE [15]	Auto-encoder with skip connections	Mean-square error	ICDAR 2013
MTRNet [23]	cGAN with auxiliary mask	Adversarial and L1	Oxford, ICDAR 2013 & 2017 MLT, CTW1500
EnsNet [28]	FCN-ResNet-18 with cGAN	multiscale regression, content, texture and total variation	ICDAR 2013, SCUT-8K
MTRNet++ [22]	cGAN with mask-refine, coarse/fine-inpaintaining branches	L1, perceptual, style and adversarial	Oxford, SCUT-8K
EraseNet [11]	Two-stage (coarse/fine) GAN-based network	adversarial, dice, reconstruction, perceptual and style	SCUT-EnsText, SCUT-8K
CTRNet [10]	Local-global content modeling with CNNs and transformer-encoder	structure, multi-scale text-aware reconstruction, perceptual, style, and adversarial	SCUT-EnsText, SCUT-8K

3   TPFNet: Our Proposed Network

Overall Idea: Our approach works by obliterating the text and painting a reasonable substitution in its place. We utilize the key idea that producing the features from a low-dimensional image is significantly simpler than producing them from a high-dimensional image. Based on this, our network operates in two parts, and the design of these two parts is shown in Figs. 1 and 3, respectively. The 512×512 input image is first resized to create a 256×256 image (T_{256}), which is fed to part 1. Part 1 of TPFNet builds a 256×256 text-free image. Part 2 of TPFNet uses this image to create a 512 × 512 text-free image. In essence, part 1 synthesizes various features, and part 2 decodes those features into a high-quality image.

We now describe the architecture and training methodology of feature synthesis (part-1) in Sects. 3.1 through 3.4 and that of image generation (part-2) in Sect. 3.5.

3.1 Encoder Design

The part-1 generator receives a 256 × 256 image. The Part 1 generator uses an encoder-decoder architecture. For the encoder, we use the pyramidal vision transformer (PVT) [25] pretrained on the ImageNet 2012 dataset. Although PVT has a higher parameter-count than some other backbone networks (refer Sect. 6), it provides unique benefits. PVT may be trained on dense image partitions to produce high output resolution, which is critical for dense prediction. PVT also uses a gradual shrinking pyramid to minimise calculations for big feature maps. These characteristics set PVT apart from ViT (vision transformer), which often produces low-resolution outputs while incurring large computational and memory overheads. PVT extends the transformer framework with a pyramid structure, allowing it to produce multi-scale feature maps for dense prediction applications. As a result, it brings together the advantages of both CNN and the transformer. Our model uses multi-scale feature maps learned by PVT to acquire the representational knowledge needed to eliminate text successfully. The output from the final part of PVT is fed into the generator's decoder, which reconstructs the images.

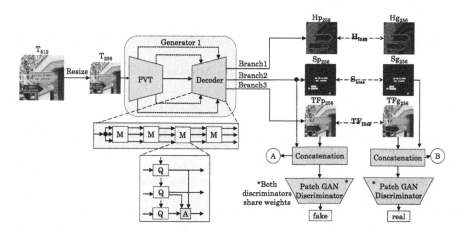

Fig. 1. Feature synthesis (part 1) in TPFNet. Ⓐ and Ⓑ connect with Part 2 (Fig. 3)

3.2 Decoder Design

Key Idea: In order to maximize the network's learning potential, we employ a multi-headed decoder structure in the generator and train it to simultaneously predict segmentation maps, high-pass filtered images, and text-free images. Our decoder integrates three branches' characteristics, providing three outcomes: a high-pass filtered image, a binary segmentation map, and a text-free image. The segmentation branch ascertains the text position, and the high-pass filter branch extracts the edges to obtain the object structure. The attention block fuses the features learned by the segmentation branch and high-pass-filter branch with those learned by the text-free image generation branch. Using three branches allows every branch to learn to forecast its own output better and to aid the other two branches by supplying extra learned representations to them. We use the Laplacian filter as the high-pass filter.

Architecture: The decoder has four identical modules, shown as "M" in Fig. 1. Module "M" has three Q-blocks and one attention block, and they are designed as follows.

Q-Block: The Q-block has two routes (refer to Fig. 2(a)). The first route incorporates a 1×1 convolution, batch normalization, and a "squeeze and excitation" block (SE Block). The SE block maps channel dependence and provides access to global information. The second route has two feature-enhancing ConvBlocks. Each block contains a 3×3 convolution layer, batch normalization, and GeLU (Gaussian error linear unit) layer.

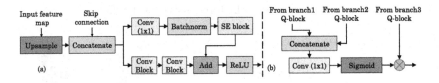

Fig. 2. (a) Q-Block, (b) Attention-Block

Attention Block: We employ an attention block, shown in Fig. 2(b), to achieve high representation ability. The attention block accepts outputs from all three Q-blocks. The attention block generates attention maps by first concatenating feature maps from branch-1 Q-block and branch-2 Q-block and then applying 1×1 convolution and sigmoid activation. Finally, it multiplies this attention map with the result of the branch-3 Q-block. The third branch of the subsequent "M" module takes its input from the attention block output. Branch 3 benefits from the knowledge acquired by branches 1 and 2. The high-pass filter enhances the feature map's edge-distinguishing capabilities, while the segmentation map pinpoints the exact location of the text.

At last, the fourth module "M" produces (1) a predicted high-pass filtered image (Hp_{256}) (2) a predicted segmentation map (Sp_{256}) and (3) a predicted text-free image (TFp_{256}). They are used to train the generator (refer Sect. 3.4).

3.3   Discriminator Design

In part 1, the discriminator accepts the concatenated vector of a text-free image and a binary mask showing text position. As a discriminator, the PatchGAN discriminator is used. This discriminator does not attempt to determine whether or not a complete image is real or fraudulent; instead, it analyses and labels small ($N \times N$) patches inside the image.

3.4   Loss Functions Used in Training

$G1(.)$ and $D1(.)$ represent the Part-1 generator and discriminator, respectively. $G1(.)$ loss is a combination of four losses: (1) loss for high-pass filtered branch (H_{loss}), (2) loss for segmentation branch (S_{loss}), (3) loss for text-free image generation branch (TF_{loss}) and (4) the conditional adversarial loss between $G1(.)$ and $D1(.)$ (GAN_{loss}). Specifically, the overall loss of $G1(.)$ is:

$$G1_{loss} = arg\ \min_{G1}\ \max_{D1}\ GAN_{loss}(G1, D1) + H_{loss}(Hg_{256}, Hp_{256})$$
$$+ S_{loss}(Sg_{256}, Sp_{256}) + TF_{loss}(TFg_{256}, TFp_{256}) \tag{1}$$

We now describe the individual loss functions. Note that the ground truth versions of the text-free image, segmentation map and high-pass filtered images are referred to as TFg_{256}, Sg_{256}, and Hg_{256}, respectively.

1. Loss for high-pass filtered branch: The H_{loss} estimated between the ground truth high-pass filtered image and the predicted one is:

$$H_{loss}(Hg_{256}, Hp_{256}) = \sum |Hg_{256} - Hp_{256}| \tag{2}$$

2. Loss for segmentation branch: Here, we use L1 loss and BCE-Dice loss. In the BCE-dice loss, "binary cross entropy" (BCE) loss computes the difference in probability distributions between two vectors. The dice loss optimizes dice-score results in over-segmented regions. Including the complementary capability of these two losses allows the model to learn more accurate segmentation masks, and the L1 loss regulates outliers. S_{loss} is expressed as:

$$S_{loss}(Sg_{256}, Sp_{256}) = \sum |Sg_{256} - Sp_{256}| - [\sum Sg_{256} \log(Sp_{256})$$
$$+ (1 - Sg_{256}) \log(1 - Sp_{256})] + (1 - \frac{2 \sum Sg_{256} \times Sp_{256}}{\sum Sg_{256}^2 + \sum Sp_{256}^2}) \tag{3}$$

3. Loss for text-free image generation branch: TF_{loss} is a combination of L1 loss and SSIM loss. The L1 loss measures the degree to which two images differ in terms of information contained within individual pixels. The SSIM loss accounts for structural details like sharp edges, color capture, and contrast characteristics. It enhances the similarity index between the ground truth image and the predicted text-free image. TF_{loss} is given as:

$$TF_{loss}(TFg_{256}, TFp_{256}) = \sum |TFg_{256} - TFp_{256}|$$
$$+ \sum (1 - SSIM(TFp_{256}, TFg_{256})) \tag{4}$$

4. Loss between $G1(.)$ **and** $D1(.)$: Since the segmented mask provides the precise location of the text, we compute GAN_{loss} with the segmented mask as the conditional variable instead of the input image. GAN_{loss} is defined as:

$$GAN_{loss}(G1, D1) = \mathbb{E}_{Sg_{256}, TFg_{256}}[\log(D1(Sg_{256}, TFg_{256}))] + \mathbb{E}_{Sp_{256}, TFp_{256}}[\log(1 - D1(Sp_{256}, TFp_{256}))] \quad (5)$$

3.5 Part-2: Image Generation

We propose an image generator $G2(.)$, which maps features generated in Part 1 to image space for generating text-free images. We employ a patch GAN image discriminator $D2(.)$ to detect whether an image is real or fake. In Part 2, the generator concatenates the predicted segmentation map (Sp_{256}) and predicted text-free image (TFp_{256}) vectors to produce the text-free image (TFp_{512}) with a resolution of 512×512. The generator also creates a text-free image (TFp_{512_o}) when provided with a vector consisting of the ground-truth segmentation map (Sg_{256}) and the ground-truth text-free image (TFg_{256}).

When $G2(.)$ takes input as a concatenated predicted segmentation mask and a predicted text-free image, it generates TFp_{512}. When $G2(.)$ takes as input the concatenated ground truth segmentation mask and ground truth text-free image, it generates TFp_{512_o}. We have used the symbol TFp_{512_o} for the ground-truth concatenated output because it is again used to generate output from generator 2.

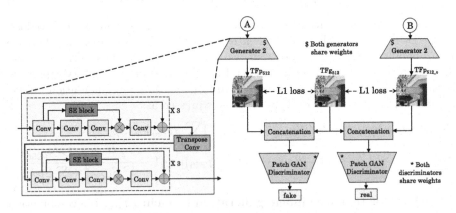

Fig. 3. Image generation (part 2) in TPFNet (Ⓐ and Ⓑ come from with Part 1 (Fig. 1)

The Part-2 generator includes the Conv block. The input is a concatenated vector, which is then fed into a series of three Conv blocks in order to extract features. The output of the first Conv block is then sent through the SE block, and the resulting vector is multiplied by the output of the third Conv block. With the help of the SE block, this draws focus to the crucial aspects. Similarly, the generator employs skip connections to maintain information flow throughout the network. In order to generate a 512×512 pixel image without any text, the network utilizes transpose convolution to upsample the feature maps, which are run through Conv blocks.

Loss Function: The loss is estimated between TFg_{512}, TFp_{512} and TFp_{512_o}. The total generator loss of part 2 is:

$$G2_{loss} = arg \min_{G2} \max_{D2} \mathbb{E}_{TFp_{512}, TFg_{512}}[\log(D1(TFp_{512}, TFg_{512}))]$$
$$+ \mathbb{E}_{TFp_{512\_o}, TFg_{512}}[\log(1 - D1(TFp_{512\_o}, TFg_{512}))] \qquad (6)$$
$$+ \sum |TFg_{512} - TFp_{512}| + \sum |TFg_{512} - TFp_{512\_o}|$$

4   Experimental Platform

Implementation Settings: We utilize PyTorch with CUDA 11.2. Experiments are conducted using two A5000 GPUs and a batch size of 32. With an initial learning rate of 1e-4, we train generators using the AdamW optimizer, D1(.) using the RMSprop optimizer, and D2(.) using the Adam optimizer. It is well known that lower-resolution images are simpler to learn from. Since the first part uses a low-resolution image as input, we use RMSprop for the first part discriminator D1(.). Since part one training can be completed more quickly than part two training, overfitting of the model occurs in part two. However, both parts must be trained concurrently. Thus, using RMSprop for D1(.) causes it to take smaller steps, whereas the use of Adam allows D2(.) in part two to take larger leaps, complementing each other's learning pace. Each part uses the cosine annealing scheduler. We use the same data and hyperparameters for all the baseline training. On 2 RTX A5000 GPUs, training TPFNet on Oxford dataset took nearly 2.3 d.

Metrics: We use PSNR, SSIM, mean square error (MSE), precision, recall, and F1-Score metrics. PSNR and SSIM help in evaluating the output image quality. For computing the last four metrics, a text detector is used on the output image [15]. A technique is effective if it achieves a near-zero score on these four metrics.

Datasets: We use the following open-source datasets for experimental purposes. They have images of size 512×512. Among the datasets, SCUT-EnsText dataset is most challenging because it is a real world dataset and includes text in different languages and fonts.

  1. The Oxford Synthetic Real Scene Text Detection [2] dataset: This dataset includes 800,000 synthetic images. We picked 95% of the data for training, 10,000 images for testing, and the rest for validation.

  2. SCUT-8K Synthetic Text Removal [28] dataset: This dataset, also called SCUT-Syn [11], has 8000 training and 800 test images.

3. SCUT-EnsText [8] **dataset:** The SCUT-EnsText has 3562 images with various text properties. We randomly choose around 70% of the images for training and the remaining images for testing to guarantee that they have the same data distribution. The training set includes 2749 images and 16460 words, while the testing set has 813 images and 4864 words.

4. ICDAR 2013 [8] **dataset:** It contains 229 training and 223 testing images with English text. Since this dataset lacks segmentation annotations, inference results on it use the model trained on Oxford dataset.

5 Experimental Results

5.1 Quantitative Results

EnsNet and MTRNet++ were reimplemented with the same settings as TPFNet by replacing our generator with EnsNet's and MTRNet++'s generators, respectively. EnsNet and MTRNet++ were trained on both the SCUT and Oxford datasets in the same manner as TPFNet. Reimplemented results are shown in the table with a (reimplemented) postfix. Other results are taken from previous papers [11,22,23].

On Oxford dataset (Table 2): TPFNet with PVT as backbone gets top PSNR and SSIM values of 44.21 and 0.989, respectively. With EfficientNetB6 as the backbone, we attain the same SSIM value of 0.989 and the second-best PSNR value of 37.9. We achieve the lowest recall, precision, MSE and F1-Score for text detection. Since MTRNet uses word/character-level ground-truth masks, it achieves least (i.e., best) precision, however, it is unfair to compare it against other techniques and hence, we have not shown its results in bold font.

Table 2. Results on the test set of Oxford Synthetic dataset (↑=higher is better, ↓=lower is better). Prec.=precision

	Method	PSNR ↑	SSIM ↑	MSE ↓	Prec. ↓	Recall ↓	F1 ↓
1.	Pix2Pix [5]	24.60	0.8970	0.54	70.03	29.34	41.35
2.	EnsNet [28]	27.42	0.9437	0.21	57.25	14.34	22.94
3.	MTRNet++ [22]	33.67	0.9843	0.05	50.43	1.35	2.63
4.	EnsNet (Reimplemented)	28.00	0.9790	0.24	57.21	14.32	24.11
5.	MTRNet++ (Reimplemented)	32.99	0.9814	0.07	50.43	2.00	3.01
6.	TPFNet (w/ EfficientNetB6)	37.90	**0.9890**	0.01	41.00	0.12	0.21
7.	TPFNet (w/ PVT)	**44.21**	**0.9890**	**0.01**	**39.00**	**0.06**	**0.17**
Results with mask							
1.	MTRNet (with mask) [23]	28.99	0.9318	0.20	35.83	0.26	0.52

On SCUT-8K dataset (Table 3): We obtain a 39.12 PSNR value and a 0.987 SSIM value with PVT as a backbone. With EfficientNetB6 as the backbone, we get 36.2 PSNR and 0.973 SSIM. CTRNet has the highest PSNR, however, TPFNet is superior on the remaining metrics. We achieve a 4.52 PSNR percentage

Table 3. Results on the test set of the SCUT-8K dataset

	Method	PSNR ↑	SSIM ↑	MSE ↓
1.	Pix2Pix [5]	25.60	0.8986	24.56
2.	STE [15]	14.68	0.4613	71.48
3.	EnsNet [28]	37.36	0.9644	0.20
4.	MTRNet++ [22]	34.60	0.9840	0.04
5.	EnsNet (Reimplemented)	36.12	0.9711	0.09
6.	MTRNet++ (Reimplemented)	33.67	0.9810	0.07
7.	EraseNet [11]	38.32	0.9767	0.02
8.	CTRNet [10]	**41.28**	0.9850	0.02
9.	TPFNet (w/ EfficientNetB6)	36.20	0.9730	**0.01**
10.	TPFNet (w/ PVT)	39.12	**0.9870**	**0.01**
Results with mask				
1.	MTRNet (w/ mask) [23]	29.71	0.9443	0.01

point difference between scores of TPFNet and MTRNet++ even though MTR-Net++ has used coarse masks. EnsNet findings are comparable to TPFNet on a small-scale dataset because of a lack of strong generalization. However, TPFNet is pre-trained on the Oxford dataset and fine-tuned on the SCUT-8K dataset, so it is challenging to overfit even on a small dataset. We achieve an MSE score of 0.01 with both EfficientNetB6 and PVT as the backbone.

On SCUT-EnsText dataset (Table 4): We obtain a PSNR value of 39.0 and an SSIM value of 0.9730 with PVT as the backbone. EraseNet has a close SSIM value of 0.9542 but a higher precision value of 53.20%, whereas TPFNet has a precision of 21.12%. Except on SSIM, we surpass state-of-the-art approaches in all the remaining metrics, showing that TPFNet's final output has greater restoration and text-removal quality.

Table 4. The results on the test set of the SCUT-EnsText dataset

	Method	PSNR ↑	SSIM ↑	MSE ↓	Prec. ↓	Recall ↓	F1 ↓
1.	Pix2Pix (Reimplemented) [5]	26.69	0.8856	0.004	69.70	35.40	47.00
2.	STE (Reimplemented) [15]	25.46	0.9014	0.005	40.90	5.90	10.20
3.	EnsNet (Reimplemented) [28]	29.53	0.9274	0.002	68.70	32.80	44.40
4.	EraseNet [11]	32.29	0.9542	0.0015	53.20	4.60	8.50
5.	CTRNet [10]	35.85	**0.9740**	0.09	40.1	1.7	**3.3**
6.	TPFNet (w/ EfficientNetB6)	37.99	0.9700	0.00082	34.88	2.18	5.66
7.	TPFNet (w/ PVT)	**39.00**	0.9730	**0.00021**	**21.12**	**1.26**	4.11

On ICDAR13 dataset (Table 5): In both no-mask and with-mask categories, TPFNet achieves best results on all metrics. Notice that on not using the mask, the performance of MTRNet remains very poor.

Table 5. Results on the ICDAR13 dataset

	Method	Eval			DetEval		
		Prec. ↓	Recall ↓	F1 ↓	Prec. ↓	Recall ↓	F1 ↓
1.	original images	70.1	81.5	75.37	70.7	81.61	75.77
2.	STE [15]	22.35	30.12	25.66	34.48	60.57	43.95
3.	Pix2Pix [5]	10.19	69.45	17.78	10.37	69.45	18.05
4.	EnsNet [28]	5.66	73.42	10.51	**5.75**	73.42	10.67
5.	MTRNet [23]	29.11	76.05	42.11	27.83	75.85	40.73
6.	TPFNet	**12.98**	**0.09**	**0.21**	12.94	**0.08**	**0.22**
Results with mask							
1.	MTRNet (w/ mask) [23]	0.18	16.67	0.36	0.18	16.67	0.36
2.	TPFNet (w/ mask)	**0.10**	**0.03**	**0.04**	**0.10**	**0.03**	**0.04**

5.2   Qualitative Results

Figure 4 shows the qualitative results of the SCUT-8K dataset. Clearly, our results are closer to the ground truth, whereas the outputs of MTRNet++ and EnsNet are blurry. Furthermore, EnsNet's outputs are incomplete and partly corrupted. EnsNet fails to completely remove text from all three images; in fact, it has also removed the plus symbol from the row (ii), which is not part of the image text. Similarly, MTRNet++ has removed some portions of the triangle from row (i), which is not part of the text; it also failed to remove some text. TPFNet is flexible enough to deal with a wide variety of challenging images, including those involving different colors (i), different fonts (ii), and different angles (iii). This shows the robustness of our network in differentiating between text and similar-looking symbols and logos.

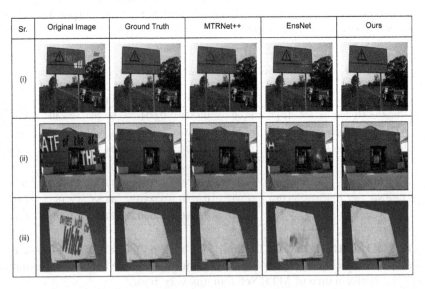

Fig. 4. Qualitative results on SCUT-8K dataset

Figure 5 shows the results for the SCUT-EnsText dataset. In row (i), MTR-Net++ and EnsNet have obliterated the text; however, they have erroneously also removed similar-looking symbols that TPFNet does not remove. It can be noticed from the row (iii) that MTRNet++ has good performance in text detection over a variety of fonts and text angles; however, it fails to differentiate between text and non-text components properly. TPFNet is effective in detecting the text and differentiating between text and non-text components in all the images.

Figure 6 shows the results of our model on real images taken from the internet. The attention mask shows that our model can successfully identify the region of interest with the text area. In all the images, our model has masked the text contained in natural images and removed the text from images without disturbing the surrounding pixels.

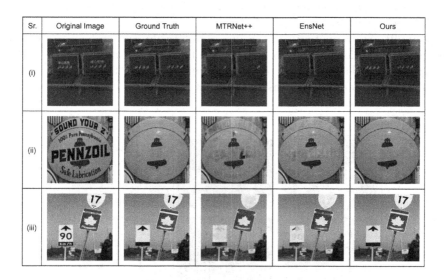

Fig. 5. Qualitative Results on SCUT-EnsText dataset

6   Ablation Study

Contribution of branches (S.No. 2 to 4 in Table 6): TPFNet has three branches: the high-pass branch, the segmentation branch, and the text-free image generation branch. On removing the high-pass or segmentation branches, PSNR degrades a lot, but SSIM reduces by a smaller amount. Thus, the high-pass branch helps detect edges, and the segmentation branch is helpful for precisely locating the text. The worst results are obtained by removing both high-pass and segmentation branches. Clearly, both branches are essential for effective edge detection with precise text removal.

Table 6. Ablation results on Scut-EnsText dataset

	Method	PSNR ↑	SSIM ↑
1.	TPFNet	39.00	0.9730
2.	Without high-pass branch	34.11	0.9640
3.	Without segmentation branch	32.11	0.9631
4.	Without high-pass and segmentation branch	30.12	0.9412
5.	Without Part 2	37.21	0.9701
6.	Without pretraning on Oxford dataset	38.23	0.9711

Contribution of part 2 (S.No. 5 in Table 6)**:** Instead of using a 256×256 image in the first part, if we directly use a 512×512 image in the first part and discard the second part, the quality metrics degrade. This confirms the usefulness of the second part.

Benefit from pretraining (S.No. 6 in Table 6)**:** We have pre-trained our model on Oxford large-scale dataset and fine-tuned it on the SCUT-8K and SCUT-EnsText datasets. There is a small impact on PSNR and SSIM when we train our network on SCUT-EnsText without pre-training it on Oxford dataset.

Fig. 6. Qualitative results on real (unseen) images taken from the internet

Contribution of loss functions (Table 7): TPFNet employs a separate loss function for each branch. Table 7 dissects the contribution of each loss function. The text-free image generation branch uses SSIM loss and L1 loss. On removing the SSIM loss, there is a small drop in PSNR but much larger drop in SSIM metric. On removing the L1 loss, the drop in SSIM metric is smaller but the drop in PSNR is higher. This is expected since PSNR measures pixel level quality and L1 loss measures pixel-level similarity. Clearly, both the loss functions are contributing meaningfully. On removing both the loss functions, only GAN-based losses are utilised and there is a large drop in the metrics.

Table 7. Loss ablation results on Scut-EnsText dataset

	Method	PSNR ↑	SSIM ↑
1.	TPFNet	39.00	0.973
2.	Without SSIM Loss	37.97	0.948
3.	Without L1 Loss	36.78	0.966
4.	Without L1 and SSIM Loss	35.88	0.929

Effect of training stretegy (Table 8): In Table 8, we assess the impact of various training strategies. When TPFNet is trained using the Pix2Pix method, it achieves a PSNR of 36.71. As input to the PatchGAN discriminator, we employ a concatenation of input image and text free image. The low PSNR achieved with Pix2Pix demonstrates the effect of concatenating a segmentation map with a text-free image as input to the PatchGAN discriminator used by TPFNet. With WGAN as our training strategy, we acquire a PSNR of 37.04. Use of WGAN leads to higher PSNR than Pix2Pix, demonstrating the influence of weight clipping. In WGAN + GP strategy, the gradient penalty is used for weight clipping and the PSNR becomes 37.78.

Table 8. Training strategy ablation results on Scut-EnsText dataset

	Method	PSNR ↑	SSIM ↑
1.	TPFNet	39.00	0.973
2.	Pix2Pix	36.71	0.959
3.	WGAN	37.04	0.964
4.	WGAN + GP	37.78	0.970

Impact of backbone and precision (Table 9): We changed the backbone from PVT to VGG16 [18], EfficientNet-B6 [20], ResNet50 [3], MobileNetV3-Large [4], and Swin-Transformer [13]. As shown in Table 9, the best results are achieved by using the Swin-Transformer as the backbone, highlighting the power of the transformer. However, it leads to a very large model size and it does not outperform PVT backbone on other datasets (results omitted); hence, we prefer PVT or EfficientNet. In Table 9, the column "full network" shows the parameters of entire network, whereas "only backbone" shows the parameters of backbone only. Note that these numbers are different from that reported in the original paper. TPFNet uses only certain number of feature-extraction (CONV or transformer encoder) layers/blocks and does not use FC layers. Hence, the parameter count reported in Table 9 is lower. On changing the precision from FP32 to FP16, PSNR reduces from 39.0 to 36.4, but the FPS increases from 14 to 20. Evidently, by changing the backbone and precision, a designer can exercise a tradeoff between model size and quality metrics.

Table 9. Backbone/precision ablation results on Scut-EnsText dataset

| | Backbone/precision | PSNR ↑ | SSIM ↑ | FPS ↑ | #Parameters | |
					Full network	Only backbone
1	With PVT	39	0.973	14	59.8	48.62
2	With VGG16 [18]	37.23	0.971	24	21.3	11.32
3	With EfficientNet-B6 [20]	37.92	0.971	19	24.4	16.02
4	With ResNet50 [3]	38.52	0.9723	17	37.9	26.76
5	With MobileNetv3-L [4]	37.3	0.973	26	12.3	3.08
6	With Swin-Transformer [13]	39.31	0.9763	9	71.8	48.78
7	With PVT and FP16 precision	36.42	0.9493	20	59.8	48.62

7   Conclusion

We proposed an end-to-end deep learning model for text removal from images. Our model outperforms previous work on nearly all configurations/metrics. Our future work will focus on the segmentation of heavily contacted characters in images taken from an oblique angle. Also, we will use unsupervised learning approaches to remove text written in unknown languages.

References

1. Bian, X., Wang, C., Quan, W., Ye, J., Zhang, X., Yan, D.M.: Scene text removal via cascaded text stroke detection and erasing. Comput. Vis. Media **8**(2), 273–287 (2022)
2. Gupta, A., Vedaldi, A., Zisserman, A.: Synthetic data for text localisation in natural images. In: Proceedings of the IEEE Conference on Computer Vision and Pattern Recognition, pp. 2315–2324 (2016)
3. He, K., Zhang, X., Ren, S., Sun, J.: Deep residual learning for image recognition. In: Proceedings of the IEEE Conference on Computer Vision and Pattern Recognition, pp. 770–778 (2016)

4. Howard, A., et al.: Searching for mobileNetV3. In: Proceedings of the IEEE/CVF International Conference on Computer Vision, pp. 1314–1324 (2019)
5. Isola, P., Zhu, J.Y., Zhou, T., Efros, A.A.: Image-to-image translation with conditional adversarial networks. In: Proceedings of the IEEE Conference on Computer Vision and Pattern Recognition, pp. 1125–1134 (2017)
6. Jo, Y., Park, J.: SC-FEGAN: face editing generative adversarial network with user's sketch and color. In: Proceedings of the IEEE/CVF International Conference on Computer Vision, pp. 1745–1753 (2019)
7. Jung, K., Kim, K.I., Jain, A.K.: Text information extraction in images and video: a survey. Pattern Recogn. **37**(5), 977–997 (2004)
8. Karatzas, D., et al.: ICDAR 2013 robust reading competition. In: 2013 12th International Conference on Document Analysis and Recognition, pp. 1484–1493 (2013). https://doi.org/10.1109/ICDAR.2013.221
9. Khodadadi, M., Behrad, A.: Text localization, extraction and inpainting in color images. In: 20th Iranian Conference on Electrical Engineering (ICEE2012), pp. 1035–1040. IEEE (2012)
10. Liu, C., et al.: Don't forget me: accurate background recovery for text removal via modeling local-global context. In: Avidan, S., Brostow, G., Cissé, M., Farinella, G.M., Hassner, T. (eds) Computer Vision - ECCV 2022. ECCV 2022. Lecture Notes in Computer Science. vol 13688. Springer, Cham (2022). https://doi.org/10.1007/978-3-031-19815-1_24
11. Liu, C., Liu, Y., Jin, L., Zhang, S., Luo, C., Wang, Y.: EraseNet: end-to-end text removal in the wild. IEEE Trans. Image Process. **29**, 8760–8775 (2020)
12. Liu, G., Reda, F.A., Shih, K.J., Wang, T.C., Tao, A., Catanzaro, B.: Image inpainting for irregular holes using partial convolutions. In: Proceedings of the European Conference on Computer Vision (ECCV), pp. 85–100 (2018)
13. Liu, Z., et al.: Swin transformer: Hierarchical vision transformer using shifted windows. In: Proceedings of the IEEE/CVF International Conference on Computer Vision, pp. 10012–10022 (2021)
14. Mao, X.J., Shen, C., Yang, Y.B.: Image restoration using convolutional autoencoders with symmetric skip connections. arXiv preprint arXiv:1606.08921 (2016)
15. Nakamura, T., Zhu, A., Yanai, K., Uchida, S.: Scene text eraser. In: 2017 14th IAPR International Conference on Document Analysis and Recognition (ICDAR). vol. 1, pp. 832–837. IEEE (2017)
16. Nazeri, K., Ng, E., Joseph, T., Qureshi, F.Z., Ebrahimi, M.: EdgeConnect: Generative image inpainting with adversarial edge learning. arXiv preprint arXiv:1901.00212 (2019)
17. Patel, C., Patel, A., Patel, D.: Optical character recognition by open source OCR tool tesseract: a case study. Int. J. Comput. Appl. **55**(10), 50–56 (2012)
18. Simonyan, K., Zisserman, A.: Very deep convolutional networks for large-scale image recognition. arXiv preprint arXiv:1409.1556 (2014)
19. Singh, S.: Optical character recognition techniques: a survey. J. Emerg. Trends Comput. Inf. Sci. **4**(6), 545–550 (2013)
20. Tan, M., Le, Q.: EfficientNet: rethinking model scaling for convolutional neural networks. In: International Conference on Machine Learning, pp. 6105–6114. PMLR (2019)
21. Tursun, O., Denman, S., Sivapalan, S., Sridharan, S., Fookes, C., Mau, S.: Component-based attention for large-scale trademark retrieval. IEEE Trans. Inf. Forensics Secur. **17**, 2350–2363 (2019)

22. Tursun, O., Denman, S., Zeng, R., Sivapalan, S., Sridharan, S., Fookes, C.: Mtr-net++: one-stage mask-based scene text eraser. Comput. Vis. Image Underst. **201**, 103066 (2020)
23. Tursun, O., Zeng, R., Denman, S., Sivapalan, S., Sridharan, S., Fookes, C.: MTR-Net: a generic scene text eraser. In: 2019 International Conference on Document Analysis and Recognition (ICDAR), pp. 39–44. IEEE (2019)
24. Wagh, P.D., Patil, D.: Text detection and removal from image using inpainting with smoothing. In: 2015 International Conference on Pervasive Computing (ICPC), pp. 1–4. IEEE (2015)
25. Wang, W., et al.: Pyramid vision transformer: a versatile backbone for dense prediction without convolutions. In: Proceedings of the IEEE/CVF International Conference on Computer Vision, pp. 568–578 (2021)
26. Yang, C., Lu, X., Lin, Z., Shechtman, E., Wang, O., Li, H.: High-resolution image inpainting using multi-scale neural patch synthesis. In: Proceedings of the IEEE Conference on Computer Vision and Pattern Recognition, pp. 6721–6729 (2017)
27. Yu, J., Lin, Z., Yang, J., Shen, X., Lu, X., Huang, T.S.: Generative image inpainting with contextual attention. In: Proceedings of the IEEE Conference on Computer Vision and Pattern Recognition, pp. 5505–5514 (2018)
28. Zhang, S., Liu, Y., Jin, L., Huang, Y., Lai, S.: EnsNet: ensconce text in the wild. In: Proceedings of the AAAI Conference on Artificial Intelligence. vol. 33, pp. 801–808 (2019)

Author Index

B
Bai, Xiang   36
Bissacco, Alessandro   3

C
Chen, Kai   22

D
Deng, En   122
Deshmukh, Gayatri   155

F
Fujii, Yasuhisa   3

G
Garcia-Bordils, Sergi   106, 137
Guo, Jie   22

H
Hua, Wei   36
Huang, Zheng   22

J
Jawahar, C. V.   137
Jia, Zhenhong   122
Jiang, Deqiang   36

K
Karatzas, Dimosthenis   106, 137
Kuang, Jianfeng   36

L
Liang, Dingkang   36
Liu, Chang   89
Liu, Yangxin   122
Lu, Yue   54
Luo, Yuchen   22

M
Ma, Cong   70
Ma, Kefan   22

M
Makwana, Dhruv   155
Mathew, Minesh   137
Mittal, Sparsh   155

P
Pal, Umapada   54

Q
Qiu, Weidong   22

R
Ren, Bo   36
Rusiñol, Marçal   106

S
Singhal, Rekha   155
Susladkar, Onkar   155

T
Teja, R. Sai Chandra   155
Tian, Jiakun   122
Tom, George   137
Tu, Mei   70
Tu, Xiao   54

W
Wang, Renshen   3
Wei, Jiajun   54

Y
Yang, Chun   89
Yang, Mingkun   36
Yin, Xu-Cheng   89

Z
Zhan, Hongjian   54
Zhang, Yaping   70
Zhao, Yang   70
Zhou, Gang   122
Zhou, Yu   70
Zong, Chengqing   70

G. A. Fink et al. (Eds.): ICDAR 2023, LNCS 14192, p. 173, 2023.
https://doi.org/10.1007/978-3-031-41731-3

Printed in the United States
by Baker & Taylor Publisher Services